Control of Cracking in Reinforced Concrete Structures

*Series Editor*
*Jacky Mazars*

# Control of Cracking in Reinforced Concrete Structures

*Research Project CEOS.fr*

Francis Barre, Philippe Bisch,
Danièle Chauvel, Jacques Cortade,
Jean-François Coste, Jean-Philippe Dubois,
Silvano Erlicher, Etienne Gallitre,
Pierre Labbé, Jacky Mazars,
Claude Rospars, Alain Sellier,
Jean-Michel Torrenti, François Toutlemonde

iSTE

WILEY

First published 2016 in Great Britain and the United States by ISTE Ltd and John Wiley & Sons, Inc.



ISTE Ltd
27-37 St George's Road
London SW19 4EU
UK

www.iste.co.uk

John Wiley & Sons, Inc.
111 River Street
Hoboken, NJ 07030
USA

www.wiley.com



Library of Congress Control Number: 2016941705

British Library Cataloguing-in-Publication Data
A CIP record for this book is available from the British Library
ISBN 978-1-78630-052-2

# Contents

**Foreword** . . . . . . . . . . . . . . . . . . . . . . . . . . . . . . . . xi

**Notations** . . . . . . . . . . . . . . . . . . . . . . . . . . . . . . . . xv

**Introduction** . . . . . . . . . . . . . . . . . . . . . . . . . . . . . . xxi

**Chapter 1. CEOS.fr Project Presentation** . . . . . . . . . . . . . . . . . . . . . 1

1.1. CEOS.fr work program . . . . . . . . . . . . . . . . . . . . . . . . 1
1.2. Testing. . . . . . . . . . . . . . . . . . . . . . . . . . . . . . . . 2
1.2.1. Tests on prismatic full-scale blocks . . . . . . . . . . . . . . . . . . 2
1.2.2. Tests on 1/3-scale beams . . . . . . . . . . . . . . . . . . . . . . 10
1.2.3. Tests on 1/3-scale shear walls . . . . . . . . . . . . . . . . . . . . 13
1.2.4. Tests on ties . . . . . . . . . . . . . . . . . . . . . . . . . . . . 17
1.3. Modeling and simulation . . . . . . . . . . . . . . . . . . . . . . . 21
1.3.1. MEFISTO research program . . . . . . . . . . . . . . . . . . . . 21
1.3.2. Benchmarks and workshops . . . . . . . . . . . . . . . . . . . . . 23
1.3.3. Numerical experiments . . . . . . . . . . . . . . . . . . . . . . . 24
1.4. Engineering. . . . . . . . . . . . . . . . . . . . . . . . . . . . . . 25
1.5. Database and specimen storage . . . . . . . . . . . . . . . . . . . . 26
1.5.1. Database CHEOPS . . . . . . . . . . . . . . . . . . . . . . . . . 26
1.5.2. Specimen storage (Renardières site). . . . . . . . . . . . . . . . . . 26

**Chapter 2. Hydration Effects of Concrete at an Early Age and the Scale Effect** . . . . . . . . . . . . . . . . . . . . . 27

2.1. Hydration effects of concrete at an early age . . . . . . . . . . . . . . 27
2.1.1. Global heating and cooling of a concrete element. . . . . . . . . . . 28

2.1.2. Differential temperature between concrete core and surface . . . . . 29
2.2. Scale effect . . . . . 32
2.2.1. Scale effect principle . . . . . 32
2.2.2. Calculating scale effect according to Weibull theory . . . . . 33
2.2.3. Worked examples of calculation with the scale effect according to the Weibull model . . . . . 37
2.2.4. Application of Weibull integral in bending and in tension . . . . . 44

**Chapter 3. Cracking of Ties** . . . . . 47

3.1. Design values and limit values . . . . . 47
3.2. Adjusting the design value for verification purposes . . . . . 47
3.3. Crack spacing equation . . . . . 48
3.3.1. Linear equation . . . . . 48
3.3.2. Relationship between the maximum spacing $S_{r,max}$ and the mean spacing $S_{r,m}$ . . . . . 49
3.3.3. Equation based on the MC2010 bond-slip relationship . . . . . 49
3.4. Equation for the mean differential strain . . . . . 50
3.5. Model accuracy when calculating the strain and crack width . . . . . 51
3.6. Example of the application of the cracking equations to a concrete tie in tension . . . . . 53

**Chapter 4. Cracking of Beams Under Mechanical Flexural Loading** . . . . . 59

4.1. Crack spacing . . . . . 59
4.2. Crack width . . . . . 60
4.2.1. Tensile stress–strain curve . . . . . 60
4.2.2. Calculating the crack width from the relative strain . . . . . 61
4.2.3. Calculating the crack width by interpolation between uncracked and fully cracked conditions (the ζ method) . . . . . 62
4.3. Examples . . . . . 69
4.3.1. Example 1: calculation of crack spacing and crack width in a thick concrete slab under heavy loads . . . . . 69
4.3.2. Example 2: calculation of crack spacing and crack width in a double thick beam . . . . . 71

**Chapter 5. Cracking in Walls** . . . . . . . . . . . . . . . . . . . . . . . . . 75

5.1. Current status of the reference texts . . . . . . . . . . . . . . . . . . . . . . 75
5.2. Validity of the physical model and calculating of the crack angle . . . . . . . . . . . . . . . . . . . . . . . . . 77
5.3. Calculation model . . . . . . . . . . . . . . . . . . . . . . . . . . . . 78
5.4. Crack spacing and slippage length . . . . . . . . . . . . . . . . . . . . . 79
5.5. Mean differential strain . . . . . . . . . . . . . . . . . . . . . . . . . 82
5.6. Calculating the crack width from reinforcing bar strains . . . . . . . . . . . . . . . . . . . . . . . . . 87
5.7. Calculating the crack width in accordance with the strut and tie model . . . . . . . . . . . . . . . . . . . . . . . . . 89
5.8. Recommendations for evaluating the cracking in walls subject to earthquake situations . . . . . . . . . . . . . . . . . . . . . . 91
5.9. Examples of application of cracking equations in a wall subjected to a shear stress in the plane of the wall . . . . . . . . . . 92
5.9.1. Example 1 . . . . . . . . . . . . . . . . . . . . . . . . . . . . 94
5.9.2. Example 2 . . . . . . . . . . . . . . . . . . . . . . . . . . . . 102
5.9.3. Example 3 . . . . . . . . . . . . . . . . . . . . . . . . . . . . 103

**Chapter 6. Minimum Reinforcement of Thick Concrete Elements** . . . . . . . . . . . . . . . . . . . . . . . . . 105

6.1. Reinforcement of reinforced concrete ties . . . . . . . . . . . . . . . . . . . 106
6.1.1. Detailed calculation from a 3D approach . . . . . . . . . . . . . . . . . 106
6.1.2. Simplified methodology for calculating concrete reinforcement . . . . . . . . . . . . . . . . . . . . . . . . . 106
6.2. Reinforcement of prestressed concrete ties . . . . . . . . . . . . . . . . . . 111
6.2.1. Crack formation in an element in tension . . . . . . . . . . . . . . . . . 112
6.2.2. Stabilized cracking stage in an element in tension . . . . . . . . . . . . 113
6.2.3. Conclusion . . . . . . . . . . . . . . . . . . . . . . . . . . . . 114
6.3. Reinforcement of beams . . . . . . . . . . . . . . . . . . . . . . . . . 115
6.3.1. Beams under monotonic mechanical loading . . . . . . . . . . . . . . . 115
6.3.2. Beams under imposed deformation and monotonic mechanical loading . . . . . . . . . . . . . . . . . . . . . . . . . 115
6.4. Reinforcement of walls . . . . . . . . . . . . . . . . . . . . . . . . . 116
6.4.1. Walls without specific requirements for cracking . . . . . . . . . . . . 116
6.4.2. Walls with specific requirements for cracking . . . . . . . . . . . . . . 117

**Chapter 7. Shrinkage, Creep and Other Concrete Properties** . . . . . 119

7.1. Introduction . . . . . . . . . . . . . . . . . . . . . . . . . . . . . . 119
7.2. Strain . . . . . . . . . . . . . . . . . . . . . . . . . . . . . . . . . 121
7.2.1. Definition . . . . . . . . . . . . . . . . . . . . . . . . . . . . 121

7.2.2. Range of applicability . . . . . . . . . . 121
7.2.3. Initial strain at loading . . . . . . . . . . 122
7.3. Shrinkage . . . . . . . . . . 124
7.3.1. Autogenous shrinkage . . . . . . . . . . 125
7.3.2. Drying shrinkage . . . . . . . . . . 125
7.4. Creep . . . . . . . . . . 129
7.4.1. Assumptions and related basic equation . . . . . . . . . . 129
7.4.2. Basic creep . . . . . . . . . . 131
7.4.3. Drying creep . . . . . . . . . . 131
7.5. Experimental identification procedures . . . . . . . . . . 133
7.5.1. Initial strain at loading time . . . . . . . . . . 134
7.5.2. Shrinkage . . . . . . . . . . 134
7.5.3. Basic creep . . . . . . . . . . 134
7.5.4. Drying creep . . . . . . . . . . 134
7.5.5. Estimation of long-term delayed strain . . . . . . . . . . 134
7.6. Temperature effects on concrete properties . . . . . . . . . . 135
7.6.1. Temperature effects on instantaneous concrete characteristics . . . . . . . . . . 136
7.6.2. Maturity . . . . . . . . . . 136
7.6.3. Thermal expansion . . . . . . . . . . 136
7.6.4. Compressive strength . . . . . . . . . . 137
7.6.5. Tensile strength . . . . . . . . . . 137
7.6.6. Fracture energy . . . . . . . . . . 138
7.6.7. Elasticity modulus . . . . . . . . . . 138
7.6.8. Temperature effects on the delayed deformations . . . . . . . . . . 139
7.6.9. Autogenous shrinkage . . . . . . . . . . 141
7.6.10. Drying shrinkage . . . . . . . . . . 141

**Chapter 8. Cracking of Beams and Walls Subject to Restrained Deformations at SLS** . . . . . . . . . . 143

8.1. Evaluation of shrinkage with bulk heating and cooling of concrete . . . . . . . . . . 144
8.2. Estimating and limiting crack widths . . . . . . . . . . 145
8.3. Estimating restraints at SLS . . . . . . . . . . 146
8.3.1. Approximate calculation of external restraint . . . . . . . . . . 146
8.3.2. Detailed calculation of a restraint on a wall . . . . . . . . . . 147
8.4. Estimation of stiffness . . . . . . . . . . 152
8.4.1. General comments . . . . . . . . . . 152
8.4.2. Simplified method . . . . . . . . . . 153
8.4.3. Principles of the detailed method . . . . . . . . . . 154
8.4.4. Worked example of a massive element thermal gradient . . . . . . . . . . 158

**Chapter 9. Effects of Various Phenomena in Combination** . . . . . . 163

9.1. Estimating crack width . . . . . . . . . . . . . . . . . . . . . . . . . . . 163
9.2. Combining effects due to imposed deformations and deformations resulting from in-service loadings . . . . . . . . . . . . . . . 164
9.2.1. Structures with water or air tightness requirements . . . . . . . . . . 165
9.2.2. Structures with durability requirements . . . . . . . . . . . . . . . . . 167
9.2.3. Minimum reinforcement . . . . . . . . . . . . . . . . . . . . . . . . . 169

**Chapter 10. Numerical Modeling: a Methodological Approach** . . . . . . . . . . . . . . . . . . . . . . . . 171

10.1. Scope . . . . . . . . . . . . . . . . . . . . . . . . . . . . . . . . . . . 171
10.2. Methodology . . . . . . . . . . . . . . . . . . . . . . . . . . . . . . . 172
10.3. Thermal and hydration effects . . . . . . . . . . . . . . . . . . . . . . 173
10.4. Drying . . . . . . . . . . . . . . . . . . . . . . . . . . . . . . . . . . 175
10.5. Mechanics . . . . . . . . . . . . . . . . . . . . . . . . . . . . . . . . 177
10.5.1. Hydration . . . . . . . . . . . . . . . . . . . . . . . . . . . . . . . 177
10.5.2. Permanent basic creep . . . . . . . . . . . . . . . . . . . . . . . . . 178
10.5.3. Reversible basic creep . . . . . . . . . . . . . . . . . . . . . . . . . 180
10.5.4. Influence of temperature on the creep velocity . . . . . . . . . . . . 181
10.5.5. Shrinkage . . . . . . . . . . . . . . . . . . . . . . . . . . . . . . . . 182
10.5.6. Drying creep . . . . . . . . . . . . . . . . . . . . . . . . . . . . . . 182
10.5.7. Steel-concrete composite modeling . . . . . . . . . . . . . . . . . . 183
10.5.8. Statistical scale effect . . . . . . . . . . . . . . . . . . . . . . . . . 184
10.6. Example simulation . . . . . . . . . . . . . . . . . . . . . . . . . . . . 185
10.6.1. Thermal and hydration simulation . . . . . . . . . . . . . . . . . . . 185

**Chapter 11. Recommendations for the use of Measurements on Mock-up Test Facilities and Structures** . . . . . . . . . . . . . . . . . . 189

11.1. General methodology of the measurements . . . . . . . . . . . . . . . . 190
11.1.1. Preliminary general approach . . . . . . . . . . . . . . . . . . . . . 192
11.1.2. Selection and choice of measuring devices . . . . . . . . . . . . . . 193
11.1.3. Method of measurement selection . . . . . . . . . . . . . . . . . . . 194
11.1.4. Measurement data-mining and analysis . . . . . . . . . . . . . . . . 194
11.2. Mock-up measurements . . . . . . . . . . . . . . . . . . . . . . . . . . 196
11.2.1. Measurement of parameters . . . . . . . . . . . . . . . . . . . . . . 197
11.2.2. Data acquisition and storage . . . . . . . . . . . . . . . . . . . . . . 207
11.3. Measurement of structures . . . . . . . . . . . . . . . . . . . . . . . . 208
11.3.1. Preliminary measurements . . . . . . . . . . . . . . . . . . . . . . . 210
11.3.2. Parameters to be measured . . . . . . . . . . . . . . . . . . . . . . . 210
11.3.3. Equipment of the measurements . . . . . . . . . . . . . . . . . . . . 211
11.3.4. Formwork . . . . . . . . . . . . . . . . . . . . . . . . . . . . . . . . 212

11.4. Example of measurement instrumentation on massive structures . . . . . 212
11.5. Example of mock-up test instrumentation . . . . . 213
11.6. Conclusion . . . . . 216

**Bibliography** . . . . . 217

**Index** . . . . . 225

# Foreword

The control of cracking in reinforced and prestressed concrete is an essential factor in ensuring the reliability and durability of structures, together with many other important properties including water-tightness and air-tightness.

Eurocode 2 (EC2) and, more recently, *fib* Model Code 2010 (MC2010) address the durability of structures and contain guidelines and rules for estimating and limiting cracking as a function of the characteristics of concrete and its reinforcement, and the exposure classifications of the works. However, these rules are normally only intended to be applied to the most common design situations. As a result, they do not take sufficient account of the behaviour of works containing massive reinforced and prestressed concrete structures, nor works which are subject to special service requirements in terms of water-tightness and air-tightness or service life and so forth. These rules are also inadequate for works requiring enhanced load protection against natural hazards or external attack. In these works, thermo-hydro-mechanical (THM) effects, scale effects and structural effects can all result in specific cracking behavior. In the case of thick rafts and walls, shrinkage and creep shall be taken into account, both in early-age concrete and in the long-term.

The purpose of this book is to provide further guidelines which can extend the existing standards and codes to cover these types of special works, especially those which are massive in nature, taking account of their specific behavior in terms of cracking and shrinkage together with other important properties such as water- and air-tightness.

The proposed rules and guidelines given in this book are based on the results of the French CEOS.fr project (*Comportement et Evaluation des Ouvrages Spéciaux – fissuration, retrait*) covering the behavior and evaluation of special reinforced concrete (RC) works with regard to cracking and shrinkage. The CEOS.fr project took place between 2008 and 2015, involving 41 French *Ministère de l'Environnement de l'Énergie et de la Mer* (MEEM), clients and project managers. The project was funded jointly by the partners of the MEEM.

The CEOS.fr project consisted partly of tests, some using full-scale solid concrete blocks and others performed on a smaller scale using laboratory models, together with the development of simulation models in collaboration with the MEFISTO project[1] under the auspices of the French *Agence Nationale pour la Recherche* (ANR). The experimental results were presented to the international scientific community and a panel of experts in these complex and rapidly changing fields assessed the simulation models. The CEOS.fr project also took account of experimental results and actual experience feedback of concrete works from the various partners.

These guidelines are addressed primarily to designers and civil engineers responsible for construction projects. Engineering rules and recommendations are illustrated at the end of each chapter using examples of design calculations, commentary on the use of models, or applicable measurement methods. Further supporting details of the basis for these guidelines may be found in the CEOS.fr test report, titled "Results obtained in the understanding of cracking phenomena" [PN 13b], which describes the results of the associated tests, the interpretation of these results and the justifications for each proposed modification to EC2 and MC2010.

The guidelines given herein reflect the latest state of the art understanding at the time of going to press. They are therefore subject to expansion and modification as new experimental data becomes available, further experience is gained and new technologies are used in future projects.

1 Maîtrise durablE de la Fissuration des InfraSTructures en bétOn.

We would like to express our sincere thanks to all those who have contributed to the publication of this document, its English version and to the, Institut pour la recherche appliquée et l'expérimentation en génie civil (IREX) for their administrative and logistical support.

Pierre LABBÉ, EDF
December 2015

# Notations

Symbols are mentioned only when they are specific and not used currently by Eurocode 2 (EC2) [NF 04, NF 06a, NF 06b] and *fib* model code 2010 (MC2010) [CEB 12]. However some symbols used less in these codes are also quoted. The units refer to the International System (IS).

| Chapter | Symbol | Description | Unit |
|---|---|---|---|
| Chapter 2 | $T_{max} - T_{ini}$ | Temperature differential at a given point between the maximum temperature reached by concrete during its setting and its initial temperature | °C |
| | $T_{adiab}$ | Adiabatic temperature | °C |
| | $\lambda$ | Thermal conductivity | $W^{-1}.m^{-1}.K^{-1}$ |
| | $Q_{\infty}$ | Hydration heat per weight unit of cement | $kJ.kg^{-1}$ |
| | $\rho_C$ | Heat capacity per weight unit of concrete | $kJ.kg^{-1}.°C^{-1}$ |
| | a | Diffusivity | $m^2.s^{-1}$ |
| | $\beta_T$ | Reduction coefficient of temperature rise calculated in accordance with adiabatic conditions: $\Delta T = \beta_T \cdot \Delta T_{adia}$ | (–) |
| | $\sigma_{cm}$ | Mean value of concrete compressive strength | MPa |
| | $f_{ctm,scale}$ | Mean value of concrete axial tensile strength taking account of scale effect | MPa |
| | $f_{ctm,scale}^{Veq}$ | Mean value of concrete axial strength taking account of scale effect, calculated according to Weibull approach | MPa |

| | | | |
|---|---|---|---|
| | $V_{ref}$ | Volume loaded by a direct tensile test that characterizes the ultimate tensile strength | $m^3$ |
| | $V_{eq}$ | Maximum volume under direct tensile strength whose failure probability is equal to the failure probability of the full scale volume | $m^3$ |
| | $k$ | Weibull exponent | – |
| | $f_{ct0,05}^{Veq}$ | 5% fractile of the tensile strength including scale effect according to Weibull approach | MPa |
| | $f_{ct0,95}^{Veq}$ | 95% fractile of the tensile strength with scale effect according to Weibull approach | MPa |
| | r | Reduction coefficient taking into account the non-uniformity of the stress field prior to the first crack | (–) |
| | $h_{ceff}$ | Effective height of the considered cross section | m |
| Chapter 3 | $\gamma$ | Ratio between the effective area and the total section area reflecting the non-uniform stresses across the section: $\gamma = \frac{A_{c,eff}}{A_c}$ | (–) |
| | $\alpha$ | Exponent of the bond–slip relationship given by Equation 6.1–1 of MC2010 | (–) |
| | $\alpha_e$ | Modular ratio $E_s/E_{cm}$ (see *EC2* and *MC2010*) | (–) |
| Chapter 4 | $\zeta$ | Parameter reflecting the crack width, depending on $\sigma_{sr}/\sigma_s$ (see *EC2* and *MC2010*) | (–) |
| | $\varepsilon_I$ | The relative strain in the section considered un–cracked | (–) |
| | $\varepsilon_{II}$ | The relative strain in the cracked section. | (–) |
| Chapter 5 | $\bar{\bar{N}}$ | Stress tensor ($N_{xx}$, $N_{xy}= N_{yx}$, $N_{yy}$) derived from structural design | MPa |
| | $N_r$ | Membrane force normal to the crack | kN/m |
| | $T_r$ | Membrane force tangential to the crack | kN/m |
| | $N_{//}$ | Membrane force normal to a plan which is perpendicular to the crack | kN/m |
| | $F_{sx}$ | Effort component on reinforcing bars along Ox | kN |
| | $F_{sy}$ | Effort component on reinforcing bars along Oy | kN |
| | $h_{eff}$ | Effective depth evaluated for the total shear wall thickness, depending on concrete cover to reinforcement | m |

| | | | |
|---|---|---|---|
| | $\theta°$ | Angle between the reinforcement in the y-direction and the direction of the principal tensile stress (see *MC2010*) | (°) |
| | $\sigma_{sx}$ | Mean steel bar stress along the x-direction | MPa |
| | $\sigma_{sy}$ | Mean steel bar stress along the y-direction | MPa |
| | $\varepsilon_{sx}$ | Mean steel bar strain along the x-direction | (–) |
| | $\varepsilon_{sy}$ | Mean steel bar strain along the y-direction | (–) |
| | $\rho_{sx,eff}$ | Percentage of steel reinforcement in the x-direction | % |
| | $\rho_{sy,eff}$ | Percentage of steel reinforcement in the y-direction | % |
| | $\alpha$ | Local distortion | rad |
| | $\rho$ | As/Ac, percentage of steel reinforcement $A_s$ based on the area $A_c$ of concrete in tension | % |
| Chapter 6 | $A_{smin}$ | Minimum cross sectional area of reinforcement | $m^2$ |
| | $h_t$ | Thickness of the concrete layer submitted in high tension to heating or cooling phase or daily temperature cycle | m |
| | $\varepsilon_c\ (t)$ | Total strain of a concrete element at time t | (–) |
| | $F_{ct,scale}$ | Concrete tensile stress at a given point along the reinforcing bar | MPa |
| Chapter 8 | $T_{max} - T_{ini}$ | See *Chapter 2* | °C |
| | $T_{min}$ | Minimum temperature up to time t | °C |
| | $\alpha$ | Free coefficient of thermal expansion of concrete | $K^{-1}$ |
| | $\varepsilon_{ca}(t)$ | Basic creep at time t | (–) |
| | $R$ | Elastic restraint factor on an infinite rigid span (CIRIA C660 guide) | (–) |
| | H, L | Height and length of a wall | m |
| | h | Distance from the given point to the base | m |
| | $K_{R0}^{i}$ | Reduction coefficient of restraint factor R | (–) |
| | $R_{bridage}$ | Restraint factor including reduction | (–) |
| | $M_{th,el}$ | Elastic bending moment induced by a restrained deformation gradient (e.g. thermal gradient) | N.m |

| | | | |
|---|---|---|---|
| | $k$ | Reduction coefficient related to the elastic bending moment | (–) |
| Chapter 10 | $\theta$ | Temperature in Kelvin | °K |
| | $\zeta$ | Hydration degree of cement | (–) |
| | $w(t)$ | Water content at time t | $Kg.m^{-3}$ |
| | $w_{\zeta\propto\infty}$ | Water quantity for cement hydration | $Kg/m^{-3}$ |
| | $D_w$ | Global coefficient of water diffusion | $m^2.s^{-1}$ |
| | $S_r$ | Degree of concrete water saturation | (–) |
| | $HR_\infty$ | Relative humidity of environment | (–) |
| | $M_w$ | Molar mass of water | $g.mol^{-1}$ |
| | $\rho_w$ | Water density | $g.m^{-3}$ |
| | $Q_\infty$ | Hydration heat per volume unit of concrete | $J.m^{-3}$ |
| | $\rho_c$ | Thermal capacity of concrete per volume unit | $J.m^{-3}$ |
| | w/c | Water–cement ratio | (–) |
| | $\Phi$ | Concrete porosity (ratio of voids to the total volume) | (–) |
| | $\zeta_0$ | Threshold of mechanical percolation (concrete change from liquid state to solid state) | (–) |
| | $S(\zeta)$ | Rigidity matrix (Hook's law) | MPa |
| | $\tau_S$ | Bond stress | MPa |
| | $\tau^M$ | Time constant of permanent creep | s |
| | $\tau^{rbc}$ | Time constant of reversible basic creep | s |
| | $\tau^{pbd}$ | Characteristic time of creep at temperature$\theta_{ref}$ | s |
| | $\Phi^{pbc}$ | Coefficient of basic permanent creep of concrete | (–) |
| | $C^c$ | Coefficient of creep stabilization | (–) |
| | $d\varepsilon^{pbc}$ | Increment of permanent basic creep | (–) |
| | $d\varepsilon^{rbc}$ | Increment of basic reversible creep | (–) |
| | $d\varepsilon^{bc}$ | Increment of basic creep | (–) |
| | $d\varepsilon^{dc}$ | Increment of intrinsic drying creep | (–) |
| | $\varepsilon^{pbc}$ | Permanent basic creep strain | (–) |
| | $\varepsilon^{rbc}$ | Basic reversible creep strain | (–) |
| | $\varepsilon^{sh}$ | Shrinkage strain | (–) |

| | | | |
|---|---|---|---|
| | $\varepsilon^{ash}$ | Autogenous shrinkage strain | (–) |
| | $\varepsilon^{kbc}$ | Potential of basic creep | (–) |
| | $\varepsilon^{pl}$ | Plastic strain | (–) |
| | $\varepsilon^{E}$ | Instantaneous elastic strain | |
| | $C^{th}$ | Reducing coefficient for characteristic times of creep according to temperature | (–) |
| | $C^{dc}$ | Reducing coefficient for characteristic time of drying creep | (–) |
| | $E_w$ | Activation energy of creep mechanisms | J/mol |
| | Gf | Fracture energy | $J/m^2$ |
| | $\Phi^{pbd}$ | Creep coefficient | (–) |
| | $h_w$ | Coefficient of water surface exchange | $m.s^{-1}$ |
| | m,n | Adjustment parameters of hydration model. Typical values: m=2.2; n=0.25 | (–) |
| | $A_w$, $B_w$ | Adjustment parameter of drying model | $m.s^{-1}$ and $m^3.kg^{-1}$ |
| Chapter 11 | α | Coefficient of thermal expansion of concrete | $K^{-1}$ |
| | w | Crack width | mm |
| | $T_{surface}$ | Temperature measured on the concrete surface | |
| | $T_{ambient}$ | Temperature measured outside of the formwork | |

# Introduction

Most concrete structures in Europe are currently designed according to Eurocode 2 (EC2). However, feedback has shown that EC2 rules do not fully reflect the complete behavior of massive concrete structures such as thick slabs or thick walls throughout time. These structures are subjected to THM effects, scale effects and structural effects that induce specific cracking patterns related to crack spacing and crack width.

To address concerns of the sustainability and durability of structures, in 2008 the French Civil Engineering Community decided to launch a joint national research project, CEOS.fr, with the aim of taking a step forward in engineering capabilities for predicting the crack pattern of special structures, mainly massive structures.

The aims of CEOS.fr project were threefold:

– to provide experimental data representative of massive test specimens;

– to develop numerical nonlinear models and damage models, to simulate concrete behavior under load and imposed deformation in accordance with the test results;

– to propose engineering rules for crack width and space assessment of possible cracking patterns in massive structures, in addition to the EC2 and *fib* Model Code (MC2010) standards.

The CEOS.fr project was carried out from 2008 up to mid-2015 around three axes: modeling and simulation in parallel with MEFISTO research project, testing on large-scale models and engineering rules.

The guidelines for the control of cracking phenomena in reinforced concrete structures are mainly dedicated to the proposed rules based on the outcomes of the CEOS.fr project and feedbacks from operated structures. These rules aim to supplement those presented in EC2 and MC2010. These guidelines were presented to a panel of experts within the framework of the Concrack4 seminar held at Ispra (Italy) in March 2014, following meetings of the EC2 Committee in charge of reviewing this standard.

The structure of this book is as follows:

Chapter 1 gives a general overview of various tests and modeling approaches, which were performed in the framework of CEOS.fr project to address the difficult topic of concrete cracking control.

Chapter 2 examines two significant effects that were identified during the massive structure tests:

– section 2.1: *hydration effects of concrete at an early age* (increase of temperature due to cement hydration followed by temperature decrease) and over time (high level of moisture retention) which lead to non-uniform concrete strains in the structural cross section of the considered element;

– section 2.2: *scale effect*, which results in the decrease of the tensile strength at an early age as observed in massive elements compared to the strength measured on laboratory test specimens, the probability to meet extreme values in the massive element volume being higher than in small size specimens.

Chapter 3 deals in particular with the 3D effect, which is characterized mainly by the non-uniform concrete stresses close to concrete cracks in the cross-section of the element. This effect is taken into account by using the $\gamma_0 \leq 1$ coefficient.

Chapter 4 proposes two methods for concrete crack width assessment in the case of massive beams or elements assimilated to massive beams.

Chapter 5 proposes an operational method for applying the rod-tie model as described in MC2010 and the calculation of the crack width derived from the reinforcing bar deformations.

Chapter 6 applies to each type of concrete elements, tie, beam and shear walls, according to whether the functioning of the concrete element is assimilated to a tie or to a shear wall.

Chapter 7 outlines the related equations presented in MC2010 and proposes some adaptations to massive elements.

Chapter 8 analyzes the relations used for the crack width calculation as given by MC2010 and proposes a method to assess the external restraints then the stiffness under the internal strains due to thermal efforts, shrinkage and creep and under external imposed deformations such as settlement.

Chapter 9 proposes an approach for calculating the concrete cracking by distinguishing the structures with waterproofing requirements from structures with sustainability requirements.

Chapter 10 describes the methodology based on the project MEFISTO results supported by the French Agence Nationale de la Recherches (ANR) and describes how to simulate the thermal and hydration effects and how to take into account those effects during the drying phase of concrete.

Chapter 11 provides recommendations on parameters measurement, measurement devices, test protocols in order to facilitate the use of measurements performed on structures, mainly on massive structures under THM effects, and the use of the feedback of experience related to this domain.

Worked examples are presented at the end of each chapter.

At the end of the book, a Bibliography gives the references of all articles referred to in the chapters.

# 1

# CEOS.fr Project Presentation

The overall objective of the CEOS.fr project is to take a significant step towards improving the engineering capabilities for assessing concrete structure crack patterns and predicting the patterns expected under anticipated design conditions. Crack control is crucial to ensure serviceability (durability and sustainability) throughout the working life of concrete structures. Current engineering practice provides some recommendations for limiting concrete cracking, with crack width and spacing control based on formulae supported by empirical data from test beams (small-sized test specimens) submitted to bending moments or tensile force. While the current codes are considered to be reasonably representative for these load cases, previous results indicate that these formulae are not fully consistent when applied to shear walls or massive structures. Hence, within the CEOS.fr national research project, several experiments on massive concrete beams were conducted to improve the knowledge of cracking phenomena, by coupling numerical modeling and experimental approaches.

## 1.1. CEOS.fr work program

The CEOS.fr program includes three work areas that are relevant for the control of cracking:

– *monotonic loads*: the purpose of which is to calibrate the available methods for predicting crack patterns and related strains under tensile or bending conditions;

– *thermo-hydro-mechanical (THM) behavior*: the purpose of which is to account for the effects on cracking of strains induced by early age behavior, shrinkage and the consequences of long-term drying, with due consideration given to boundary conditions;

– *seismic and cyclic loads*: the purpose of which is to consider the seismically induced crack patterns in shear walls, either during or after the seismic event, taking into account the cumulative damaging effects of all loads.

The three aforementioned work areas were studied according to the following three approaches:

– *testing*: implementing tests on large-scale specimens (full scale, 1/3 scale ties and beams, 1/3 scale shear walls) with well-identified boundary conditions and accurate crack pattern monitoring;

– *modeling*: applying the existing numerical models and developing specific models compatible with engineering issues;

– *engineering*: developing design rules from test results and numerical simulations with the aim of establishing *guidelines for the control of cracking phenomena in reinforced concrete structures.*

## 1.2. Testing

Few experimental results which relate to large or massive structures are found in the literature. Hence, four types of specific tests were performed under the framework of the CEOS.fr project, as described in the following section.

All test block specimens were comprehensively monitored to locate and follow crack propagation and to measure crack spacing and crack widths. Digital Image Correlation (DIC) was specifically used to measure crack widths on the whole surface of the test specimens (see section 11.5 for details).

### 1.2.1. *Tests on prismatic full-scale blocks*

#### 1.2.1.1. *Tests on free strain blocks*

Figure 1.1 depicts the basis of the seven full-scale blocks (dimensions 6.10 m × 1.60 m × 0.80 m), designated RL1 to RL7. These blocks were constructed according to specifications summarized in Table 1.1. Two identical reference blocks, RL1 and RL6, were constructed from C50/60 class concrete and 16 HA Ø 32 reinforcing bars. The other individual blocks were produced using the same criteria, but with one feature changed compared to the reference blocks, either concrete grade, concrete cover, reinforcing bar diameter or reinforcement ratio (Table 1.1). The average reinforcement ratio was approximately 1%, which is representative of such structures.

**Figure 1.1.** *RL block with free deformations (Top) and its reinforcement (Bottom)*

| Block/Beam | Specificity | Concrete cover (mm) | Cement | Concrete class | Reinforcement bars HA (top) |
|---|---|---|---|---|---|
| RL1 RL6 | Reference beams | 50 | CEM1 52,5N CE CP2 NF | C50/60 | 16 Ø 32 |
| RL2 | Minimum percentage of reinforcement | 50 | CEM1 52,5N CE CP2 NF | C50/60 | 4 Ø 32 |
| RL3 | Increased bar diameter | 50 | CEM1 52,5N CE CP2 NF | C50/60 | 10 Ø 40 |
| RL4 | Increased concrete cover | 70 | CEM1 52,5N CE CP2 NF | C50/60 | Ø 32 |
| RL5 | Reduced concrete resistance | 50 | CEM1 52,5N CE CP2 NF | C30/37 | 16 Ø 32 |
| RL7 | Addition of two inclusion vertical cables | 50 | CEM1 52,5N CE CP2 NF | C50/60 | 16 Ø 32 |

**Table 1.1.** *Characteristics of RL blocks*

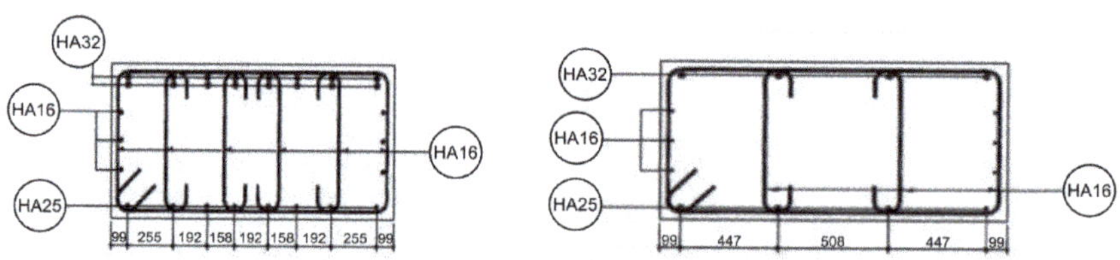


**Figure 1.2.** *Reinforcement bars HA of block cross-sections: RL1, RL5, RL6 and RL7 (left figure) and RL2 (right figure)*

Post casting, the seven blocks were freely matured for a period of at least 4 weeks, with limited protection provided against major weather conditions. Following this period, each block was moved onto the test bench and secured with four prestressing bars used to hold the block in place. The block was then submitted to a monotonic bending load by two rows of 4 × 1,000 kN capacity jacks, 6,000 mm spaced and symmetrically positioned under the central part of the beam (Figure 1.3).

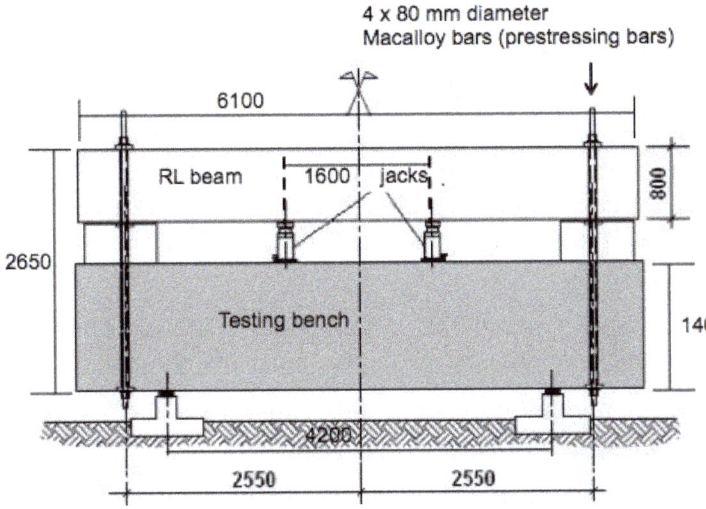


**Figure 1.3.** *Specification and principle of bending test for RL block (beam)*

Flexural tests were then performed at incremental steps of 150 kN, up to a maximum load varying from 2,000 to 2,500 kN depending on the beam. The maximum bending moment applied varied from 1,600 to 2,000 kNm, except for RL2. This approach was taken to ensure that the crack pattern was completely stabilized for each beam and, hence, that the Service Limit State (SLS) is fully addressed.

The top reinforcement bars support a maximum stress of approximately 400 MPa (EC2 and MC2010), which is the theoretical steel stress at the concrete crack.

With the exception of RL2, which was close to failure, resulting from the minimum percentage of reinforcement used, block failure did not occur.

To confirm crack pattern evolution with bending load, specific DIC procedures were developed, which provided full measurement of the crack width evolution and crack spacing with the applied bending load.

The resultant DIC surface measurements were correlated with internal measurements of the temperature evolution inside the specimen blocks (temperature gradient), overall and local strain evolution and rebar stress. Hence, to support this approach, at least three surfaces are instrumented, positioned centrally in each block. The analysis of the resultant data has led to a greater understanding of the cracking phenomena associated with the most significant cracks (see Figure 1.4):

– the crack pattern is largely dependent on the early age of the block. In massive blocks, high temperature levels are reached in the core. Temperature gradients can lead to early cracking on the block surface after formwork removal or at the block core with restrained shrinkage due to the reinforcement layers;

– surface cracks generally close during the maturing phase and do not influence the crack pattern obtained under mechanical loading;

– core cracks may influence the crack pattern obtained under mechanical loading as crack spacing and crack widths are larger;

– for massive structures, where reinforcement bar cover is more significant, the concrete crack widths are generally greater. Section 4.2 presents an approach which gives more details on how this test feature is taken into account.

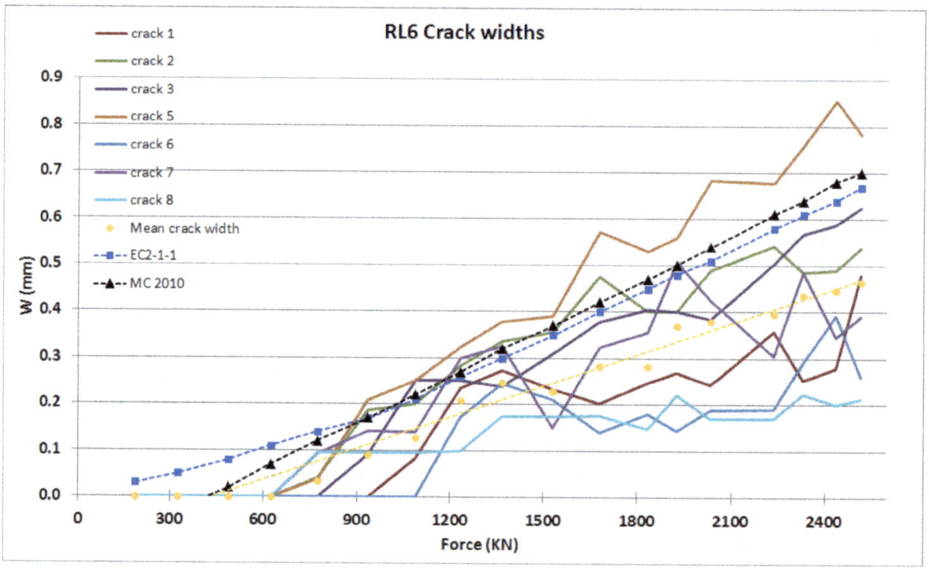


**Figure 1.4.** *Crack evolution on prismatic beams tested in bending: a comparison between the test results from several cracks (solid lines) and the comparable code-based previsions (dashed lines) for full scale beam RL6*

In addition, tensile resistance of thick elements is lower than the tensile strength calculated in accordance with EC2-1-1 ($f_{ct,\ 0.05} = 0.7\, f_{ctm}$).

$$f_{ct,\ 0.05} \approx 0.5\, f_{ctm}$$

This relationship is due to the 3D stress distribution in the thick element cross-section and the associated scale effect, since the cross-section of the massive beam is significantly larger than the cross-section of the specimens tested in the laboratory, used to calibrate the EC2-1 formula. Hence, the actual $f_{ct,\ 0.05}$ value is more likely to be reached in a massive structure.

NOTE.– Section 11.5 of the Guidelines gives more details on the measurement system used for the test specimen blocks.

#### 1.2.1.2. *Tests on blocks subjected to retrained shrinkage*

The three concrete blocks RG8, RG9 and RG10 were constructed using three parts (Figure 1.5):

– a central prismatic block (6 m × 0.50 m × 0.80 m), which comprises the block;

– two head blocks (0.9 m × 2.2 m × 0.9 m);

– two steel struts, which restrain strains to induce cracks in the block.

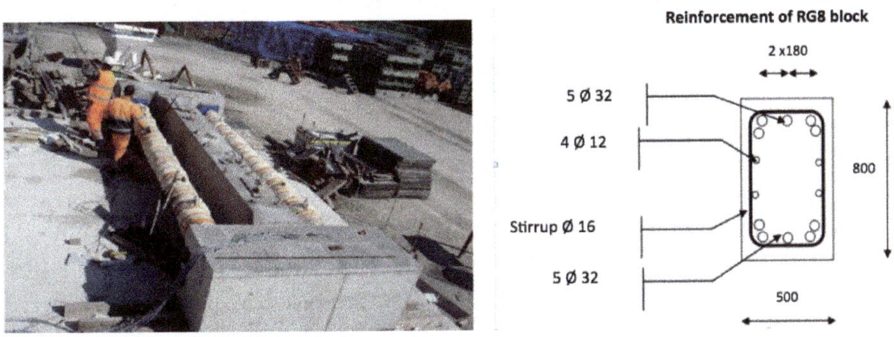


**Figure 1.5.** *RG8 block with restrained shrinkage (I-shaped beam and steel struts)*

During the RG block-maturing phase, the following were measured: overall strains, local strains and temperatures on the block surface, the block core and along the rebar. Average strut forces were also measured during the test. The RG block local concrete strains were compared with the strains measured in the concrete specimens cast on site (freely matured test specimens and quasi-adiabatic test specimens) to assess the restrained concrete shrinkage.

| Block/Beam | Specifics | Cover (mm) | Cement | Concrete class | Reinforcement ratio |
|---|---|---|---|---|---|
| RG8 | Reference beam | 50 | CEM I 52,5N CE CP2 NF | C50/60 | 2% |
| RG9 | Reduced reinforcement | 50 | CEM I 52,5N CE CP2 NF | C50/60 | 0.6% |
| RG10 | Increased cover | 70 | CEM I 52,5N CE CP2 NF | C50/60 | 2% |

**Table 1.2.** *RG block characteristics*

These changes demonstrate early age cracking due to THM effects, which result from the temperature increase during the concrete setting, the temperature decrease and the restrained shrinkage. During the tests, of a 400-hour duration, three to four cracks appeared in the central beam sections, despite the fact that the stabilized cracking stage was never reached. Cracks were detected by the measurement system (strain gauges, Linear Variable Differential Transformer (LVDT) and fiber sensors).

The main difficulty in carrying out this test is to forecast where cracks will occur in order to implement the sensors at the right place. However, the transfer of shear forces between concrete and steel bars in the vicinity of cracks can be derived from measurements.

The maximum and minimum crack width tests result from RG8 (with 2% reinforcement), maximum crack width 107 μm, was less than that obtained from RG9 (with 0.6% reinforcement), maximum crack width 126 μm.

#### 1.2.1.3. *Main experimental outputs of massive test blocks*

The most significant finding from the CEOS.fr project is an improved understanding of the THM effects for early age concrete. At an early age, the massive elements are submitted to non-uniform strains, which may induce cracking at this stage. This effect is unavoidable in practice and is generated by:

– temperature gradients between the core and the surface of the massive element;

– internal strains generated due to the temperature profile, shrinkage and creep.

In addition, the assessment and interpretation of the test results for massive elements has improved the understanding of the influence of two phenomena: 3D effects, as the massive elements are always submitted to 3D non-uniform strains, and probabilistic scale effects.

To explain the effects seen on massive structures, it is necessary to assume that the mean tensile strength is reduced compared with the split test and that the stress distribution is non-uniform, mainly due to the probabilistic scale effect (referred to in MC2010 clause 7.12.5.4): the volume of concrete submitted to high tensile stress is larger when compared to that submitted to tensile stresses in a specimen under tensile test (see NF EN 12390-6 split tensile test). In this example, the likelihood that the tensile strength value $f_{ct,0.5}$ is reached for massive structures is much greater than that in a laboratory test specimen.

The probabilistic scale effect can be simulated by using the Weibull theory, and the weak value of the tensile strength can be mainly explained by the use of a simplified approach based on this theory (see section 2.2).

**Figure 1.6.** *RL1- 1/3 scale beam under reinforcement*

### 1.2.2. *Tests on 1/3-scale beams*

Six 1/3-scale blocks (1.90 m × 0.25 m × 0.44 m) were manufactured in accordance with the specifications given in Table 1.3.

| Beam1/3 | Specifics | Cover (mm) | Concrete class | Reinforcement bars (top) | Stirrup spacing (mm) |
|---|---|---|---|---|---|
| A | Standard concrete | 15 | C 50/60 | 16 Ø 10 | 100 |
| B1 – B2 | First use of micro concrete B | 15 | C 50/60 micro concrete | 16 Ø 10 | 100 |
| C | Reduced number of rebars | 15 | C 50/60 micro concrete | 8 Ø 14 | 100 |
| D1 –D2 | Stirrup spacing increased | 15 | C 50/60 micro concrete | 16 Ø 10 | 144 |

**Table 1.3.** *1/3-scale block characteristics*

The purpose of these tests was to highlight the scale effect on the behavior of the beam when compared with the full-scale prismatic blocks, and to verify its validity for structures designed in accordance with EC2, which relates to the crack width and spacing assessment.

Each 1/3-scale beam, with free deformations, is subjected to monotonic loading adapted from the full-scale beam flexural tests. These tests were conducted in a step-by-step manner. Each of the 18 load steps is applied by two jacks at a spacing of 500 mm, up to a maximum load of 234 kN (Figure 1.6). In practice, a pre-load is applied to the 1/3-scale beams to calibrate the load system. Each load step was scaled according to the crack width measurement and crack spacing.

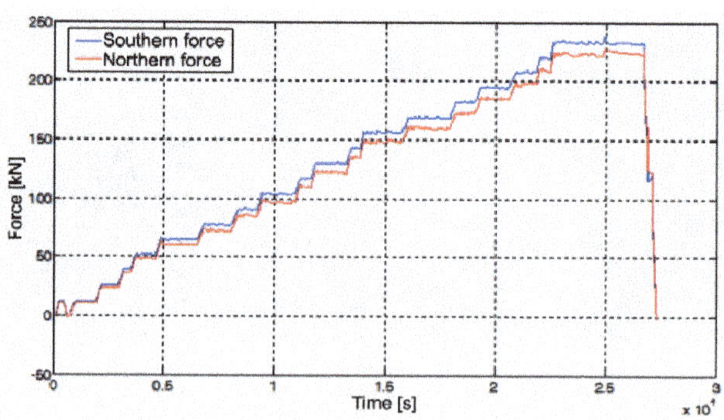


**Figure 1.7.** *1/3 scale beams load path*

| *Measurement* | *Sensor* | *Number* | *U.M.* | *Location* | *Comments* |
|---|---|---|---|---|---|
| *Temperature* | PT 100 | 2 | °C | Concrete (internal) | Embedded |
| *Force* | Transducer | 2 | kN | Central | Upward vertical direction |
| *Strain (+ temperature)* | Vibrating wire strain gauge (VWSG) | 4 | µdef | Concrete (internal) | Longitudinal direction |
| *Strain* | Gauge | 3 | µdef | Upper edge | Longitudinal direction |
| *Strain* | Sensor with Fiber Bragg Grating (FBG) | 4 | µdef | Welded on rebars | Longitudinal direction |
| *Displacement* | Long base optical extensometer (LVDT) | 8 | mm | Upper and lower edge | Vertical direction |

**Table 1.4.** *1/3-scale beam test measurements*

Figure 1.8 gives the location of the sensors on the 1/3 scale beam B2: temperature probes PT 100, strain gauges glued to the upper level rebars and Vibrating Wire Strain Gauge (VWSG) embedded within the concrete.

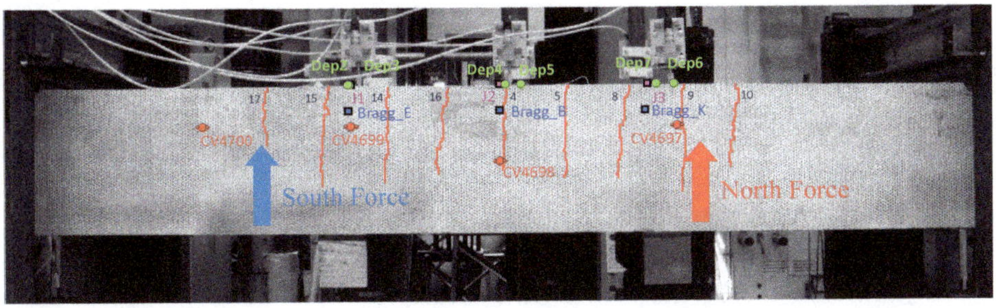


**Figure 1.8.** *Sensor location on the painted side of the 1/3 scale tested beam and crack pattern detected at the test final load using digital image correlation (Fiber Bragg Grating (FBG)), Vibrating Wire Strain Gauge (quoted CV), temperature sensors (Pt 100), displacement sensors (Dep)*

For 1/3-scale beams, the test results demonstrate that the mean crack spacing and widths at the stabilized cracking stage are consistent with the crack spacing and width calculated using the MC2010 formulae. Note that the beam size used corresponds to test results and on-site observations used to establish the EC2 and MC2010.

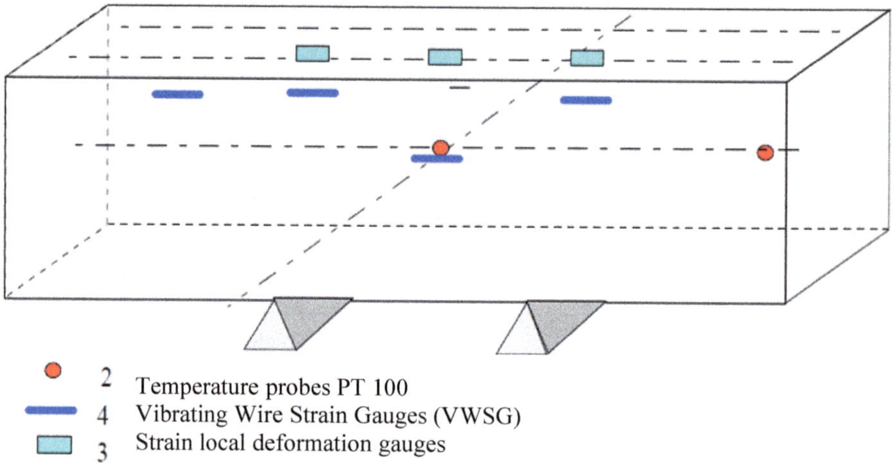


**Figure 1.9.** *B2 tested 1/3-scale beam displaying the location of various sensors*

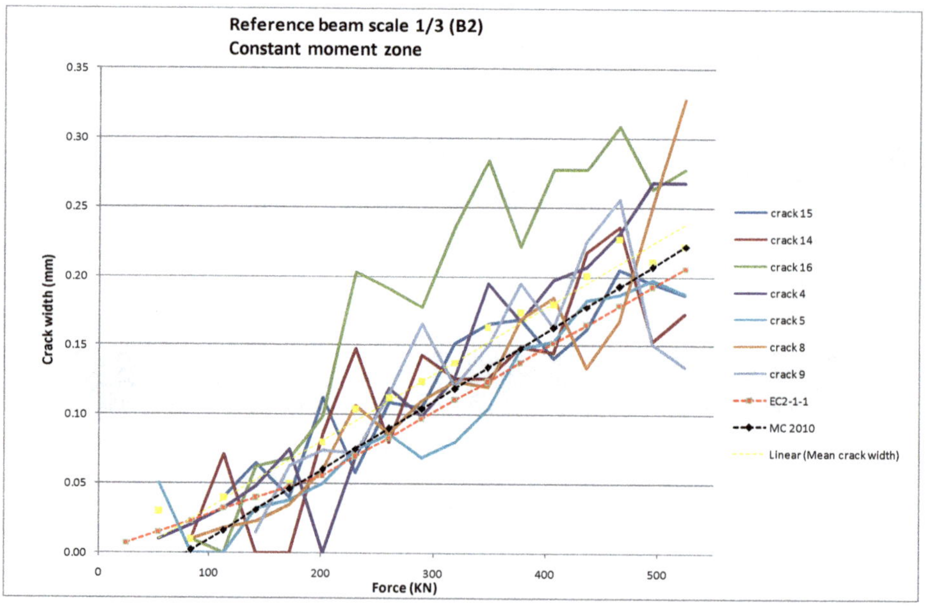


**Figure 1.10.** *Crack evolution on prismatic beams tested in bending: a comparison between the test results from several cracks (solid lines) and the comparable code-based previsions (dashed lines) for a 1/3-scale beam*

### 1.2.3. *Tests on 1/3-scale shear walls*

Four 1/3-scale shear walls (height = 1.05 m, length = 4.2 m, thickness = 0.15 m), designated SHW1 to SHW4, were manufactured according to the specifications summarized in Table 1.5. In addition to the similitude rules, the design of these test specimens was driven by two conditions:

– to accurately reproduce reinforced thick shear walls representative of industrial structures;

– to adapt the availability of the testing means, the capacity of jacks of the testing bench being limited to 4,500 kN.

As presented in Table 1.5, the walls differ from each other due to either the type of concrete or the reinforcement mesh.

The mock-up was installed in a rigid steel frame, thereby avoiding any large reactions on the test slab specimen. This also enables improved control of applied forces and boundary conditions. The load actuator was installed between the upper beam and the steel frame.

| Shear wall | Concrete class | Reinforcement bars | Load type |
|---|---|---|---|
| SHW1 | C25/30 | HA10 – 100 mm × 100 mm | Reversing cyclic loading |
| SHW2 | C40/50 | HA10 – 100 mm ×100 mm | Reversing cyclic loading |
| SHW3 reference wall | C40/50 | HA10 – 100 mm × 100 mm | Non-reversing loading |
| SHW4 | C40/50 | HA8 – 80 mm × 80 mm | Reversing cyclic loading |

**Table 1.5.** *Characteristics of 1/3 scale shear walls*

To obtain the most even shear force spreading possible, two highly reinforced horizontal beams were horizontally connected to the top and bottom of the shear wall. As the walls were designed without flanges, vertical steel bars were added to reinforce both edges to control the crack opening due to the induced bending effect (Figure 1.10).

The shear walls given in Table 1.5 were tested incrementally until failure. Three of the walls were subjected to horizontal reversing load applied in a series of three cycles, with a ±300 kN increasing load step. The SHW3 reference wall specimen was tested under a non-reversing load applied in a single "push" direction, as shown in Figure 1.10.

**Figure 1.11.** *Shear wall specimen SHW3 (left figure) and shear wall reinforcement (right figure)*

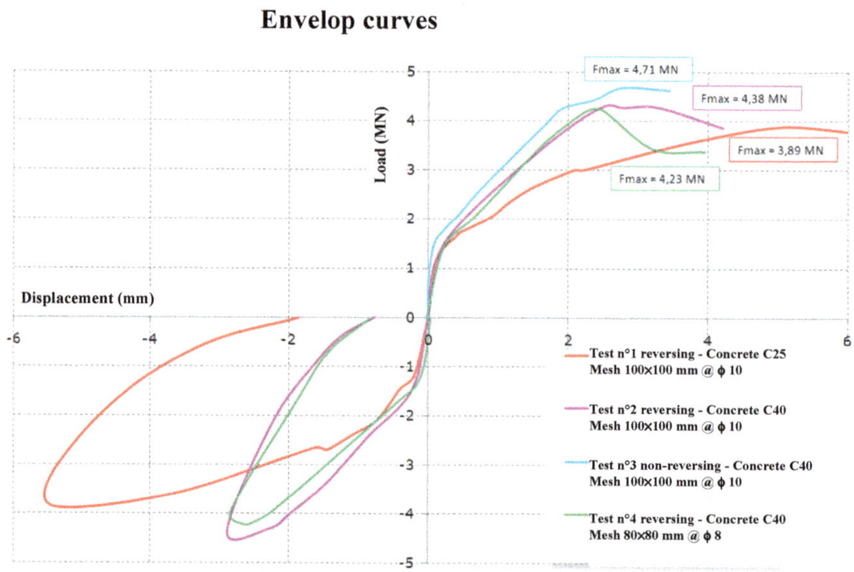


**Figure 1.12.** *Comparison of force-displacement envelope curves related to the four SHW tests*

During these tests, the ultimate strength is not reached through rupture in the central zone, but by sliding along the plane between the wall and the upper beam, extended by an inclined plane joining the horizontal bottom support. This effect has been reproduced across all the four walls.

#### 1.2.3.1. *Non-reversing loading test*

During the non-reversing loading test, the influence of the following factors has been assessed, with the spacing $S_r$ between cracks compared to EC2 and MC2010-formulae-based results. In addition to this approach, other results from the SAFE experiment (ISPRA) have been re-analyzed, especially at Ultimate Limit State (ULS), with the CEOS.fr outcomes demonstrating consistent agreement with the SAFE results.

##### 1.2.3.1.1. Concrete cover

The results of the shear wall tests show that it is advisable to use, for calculation purposes, either the cover for each rebar layer or the mean value of the two covers in both directions. Of the two approaches, the second (mean value) solution is considered to be simpler to apply.

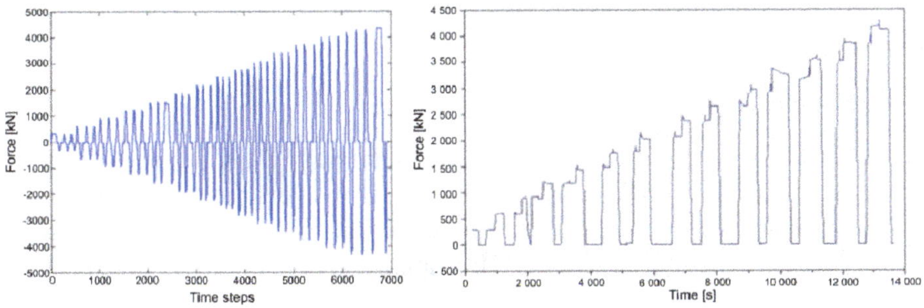


**Figure 1.13.** *Load path of specimens SHW1, SHW2, SHW4 (left) and SHW3 (right)*

##### 1.2.3.1.2. Angle θ

Angle $\theta$ is the angle between the reinforcement in the horizontal direction and the direction of the crack determined by the principal tensile stress. If the reinforcement layers are in accordance with the optimized ratio between horizontal and vertical reinforcement, an error in angle $\theta$ assessment does not result in a significant error in the assessment of crack spacing and crack width.

#### 1.2.3.1.3. Structural tensile resistance of concrete $f_{ctm}$

The test results show that the mean structural tensile resistance $f_{ctm}$ is 40% less than the theoretical value of $f_{ctm}$ when the first crack occurs. This reduction is due to the 3D effects for massive elements, resulting in stress variations across the section.

Scale effects between $f_{ctm}$ measured on specimens under laboratory conditions and $f_{ctm}$ results observed on the large concrete volume of the tested shear walls are also a factor which contributes to reducing $f_{ctm}$ in shear wall tests. In addition to these scale effects, the effect of concrete struts working in compression may also decrease the tensile strength.

The value of the compressive strength $f_{cm}$ demonstrates some impact (approximately 20% between SHW1 and SHW2 shear wall tests n°1 and 2, respectively) on $S_r$ value and cannot be discounted, although it should be noted that both EC2-1 and MC2010 do not take this parameter into account.

The crack width $w_d$ has been measured for comparison with EC2 and MC2010 formulae.

#### 1.2.3.1.4. Influence of reinforcement ratio

For test SHW4, the overall reinforcement ratio falls from 1% to 0.8%, while $S_r$ is seen to increase by up to 22%. However, the application of the EC2 and MC2010 formulae gives a reduction in $S_r$, due to the formulae being based on the effective area. This approach appears to be unsuitable for the calculation of $S_r$ (Figure 1.14).

### 1.2.3.2. *Comparison of reversing and non-reversing loading tests*

$\theta$ varies by approximately 10% depending on the "pushing" direction applied during the test, and is not dependent on $f_{ct}$.

The test results show that cracks close during the reversing load test, when the load is approximately zero, and that cracks do not close during non-reversing load tests. Crack width increases with the load during the cyclic testing.

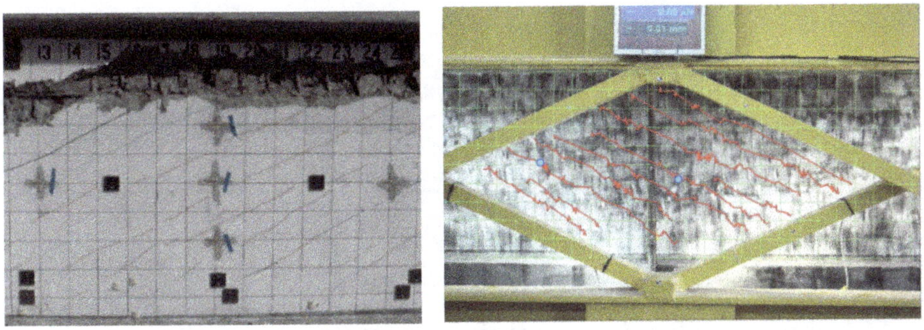

**Figure 1.14.** *Damaged central part of shear wall SHW3 at collapse, under non-reversing loading (left figure) – reconstruction of crack pattern with DIC on the opposite side (right figure)*

## 1.2.4. *Tests on ties*

Nine large ties, constructed from concrete C40/50 in accordance with the requirements given in the NF EN 206-1 standard, were tested under laboratory conditions with various sizes and types of reinforcement used as described in Table 1.6.

| Type of tie | Number of ties | Dimensions | Reinforcement |
|---|---|---|---|
|  | 2 | 135 × 135 × 3,200 mm | 1 rebar Φ25 |
| 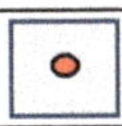 | 2 | 170 × 170 × 3,00 mm | 1 rebar Φ40 |
| 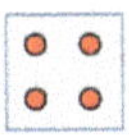 | 3 | 355 × 355 × 3,200 mm | 4 rebars Φ25<br>One tie with stirrups<br>Φ10@200 mm |
| 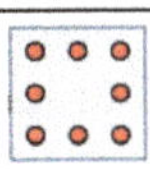 | 2 | 355 × 355 × 3,200 mm | 8 rebars Φ16<br>with cross bracing Φ8 |

**Table 1.6.** *Tie characteristics*

### 1.2.4.1. *One rebar concrete tie test*

For each bar diameter, two types of loading were applied:

– direct tension loading: concrete ties were loaded at a slow speed, with the crack pattern recorded for each loading stage ($R_e/4$, $R_e/2$, $3/4R_e$ and $R_e$, elastic limit). Tests were performed up to the maximum rebar extension allowed by the stroke of the jacks monitoring the loading;

– cyclic tension loading: these tests were performed using the same methodology as for the direct tension tests. In addition, four loading–unloading cycles were performed at each of the two post-yielding stages, corresponding to bar elongations equal to 25 mm and 50 mm, respectively.

#### 1.2.4.2. *Four or eight rebar concrete tie tests*

Only direct tensile tests were performed for ties with four or eight rebars, under the same conditions as described in section 1.2.4.1.

#### 1.2.4.3. *Measurements*

The bar and concrete elongations of the tie were recorded with sensors installed at both ends of the specimen. The crack pattern (crack location and width) was recorded at each loading stage. For the latter, two types of crack measurement apparatus were used: micro-sensors and a microscope linked to a screen [MER 14].

#### 1.2.4.4. *Main outputs from tie test results*

*Tensile strength*

The results from tie tests demonstrate that the first cracks appear in the concrete cross-section under a lower tension load than predicted by the conventional tensile stress $0.7 f_{ctj}$: e.g. values of 1 or 2 MPa compared to the 3 MPa predicted by the formulae and the even higher values measured on cylinders.

*Crack spacing*

The maximum spacing, $S_{r,\max}$, measured between stabilized cracks is less than that predicted by the EC2 – expression [7.11] applied to large ties, which significantly overestimates this spacing. The maximum spacing calculated from the MC2010 – expression 7.6.4 appears to provide spacing values with a better agreement with the test results, although still higher, as given in Table 1.7.

Test results related to crack spacing measured on tie specimens have been integrated into the statistical study giving the crack spacing for a tie in tension (see section 3.3).

| | | | Experimental | | | Theoretical according to MC2010 and EC2-1 | |
|---|---|---|---|---|---|---|---|
| Tie | Diameter rebar (mm) | Number of rebars | Average crack spacing (mm) | Standard deviation crack spacing (mm) | Characteristic crack spacing (mm) | 2ls, max MC2010 (mm) | *Sr*, max EC2-1 (mm) |
| 4 | 40 | 1 | 178 | 67 | 289 | 392 | 426 |
| 5 | 40 | 1 | 178 | 65 | 284 | 392 | 426 |
| 9 | 25 | 1 | 168 | 58 | 264 | 360 | 430 |
| 10 | 25 | 1 | 152 | 53 | 239 | 360 | 430 |
| 1 | 25 | 4 | 200 | 80 | 331 | 590 | 664 |
| 2 | 25 | 4 | 200 | 78 | 329 | 590 | 664 |
| 3 | 25 | 4 | 246 | 77 | 373 | 590 | 664 |
| 4 | 16 | 8 | 160 | 59 | 257 | 437 | 534 |
| 5 | 16 | 8 | 188 | 79 | 318 | 437 | 534 |

**Table 1.7.** *Comparison of crack spacing between test measurements and values calculated from EC2 and MC2010*

### *Crack width*

The maximum crack width measured is close to the value calculated both from EC2-1 and MC2010 (Figures 1.15 and 1.16), which suggests that the relative mean strain value might be underestimated by these codes.

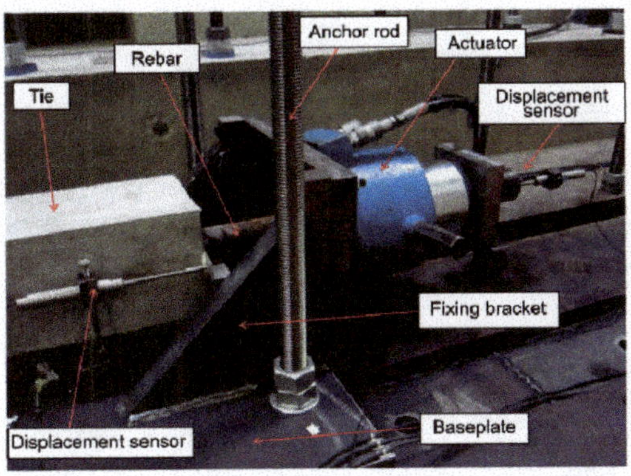


**Figure 1.15.** *Jack and displacement sensor*

**Figure 1.16.** *Tie with eight reinforcement bars at crack saturation stage*

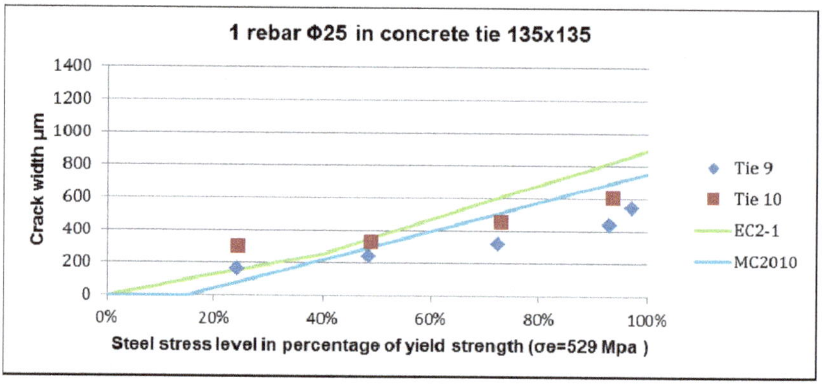


**Figure 1.17.** *Crack widths compared between test results and EC2/MC2010 calculations for one rebar tested ties 9 and 10*

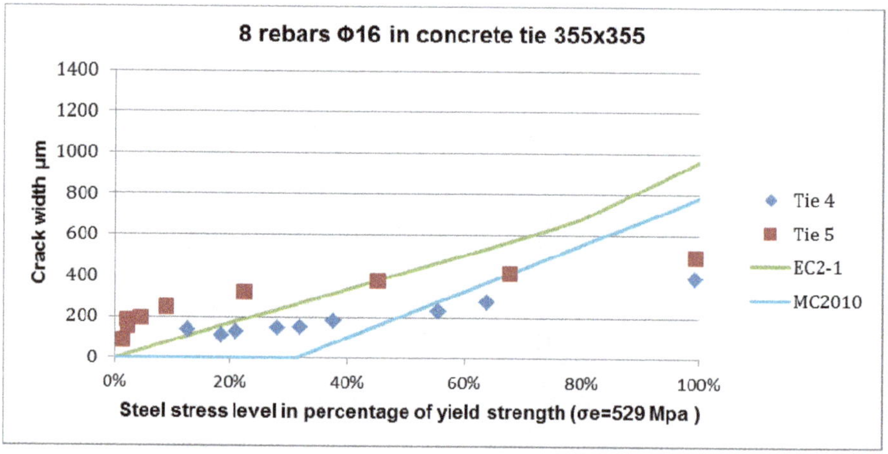


**Figure 1.18.** *Crack widths compared between test results and EC2/MC2010 calculations for eight rebar tested ties 4 and 5*

## 1.3. Modeling and simulation

Numerical modeling played a key role in the interpretation of test result data obtained from the cracking of massive elements at an early age and 1/3 scale shear wall tests. Comparison of model numerical results with the experimental data obtained from instrumentation has provided an improved understanding of the physical phenomena due to scale effects and THM effects in massive structures [BUF 16]. As a result, it was possible to use modeling and other simulation tools to extend the range of the experimental program results, through numerical "virtual" modeling of the physical processes.

A considerable number of studies have been performed using models and modeling. Several actions were launched in parallel: benchmarking of the basis of existing experimental results, either from the CEOS.fr project experiments or the established dedicated research programs, such as MEFISTO (see section 1.3.1).

These benchmarking actions helped as a first phase to test the existing models and as a second phase to improve the performance of these models or to develop new operational models for the analysis of the experiments performed. Finally, in addition to these "physical" experiments, the models developed were used to carry out a program of "numerical" experiments, which provides a comprehensive database, allowing engineers and scientists to make proposals for improving or renewing standards.

### 1.3.1. *MEFISTO research program*

In parallel with the CEOS.fr project, the Agence National pour la Recherche (ANR) launched the MEFISTO research project in 2008. As a result of the collaboration between the researchers' teams of MEFISTO and CEOS.fr projects, it was possible to develop models, which consider the following topics and approaches:

– modeling of effects under monotonic loading in connection with the overall performance of the material (stress-strain model) and the local damage process (trajectory and width of cracks);

– modeling of THM-coupled effects on concrete at an early age and assessment of induced stresses and local damage.

The MEFISTO project also provided support to the CEOS.fr project for the design and construction of the full and 1/3-scale test bodies.

The development of numerical tools has provided an improved assessment of the macro-crack positions and crack openings, using post-processing tools for finite-element-based simulations (2D or 3D) and simplified approaches, such as multi-fiber beam models (based on 1D models).

### *Discrete and mixed methods*

Due to the joint use of continuous and discrete approaches, mixed approaches allow the simulation of the cracking pattern at the center of the model. The challenge posed by this approach is to allow the two models to coexist in the same calculation processing in parallel.

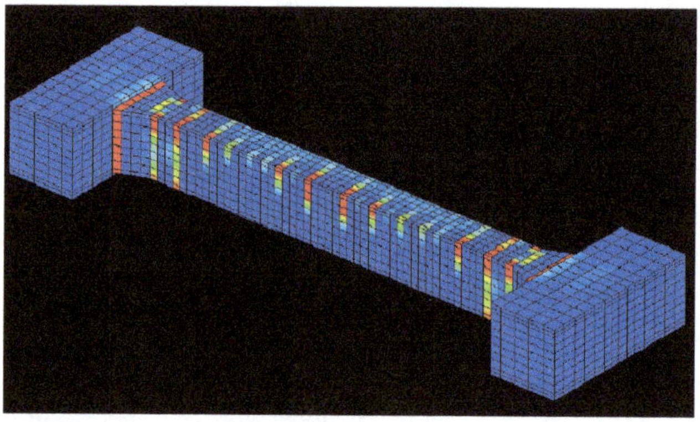

**Figure 1.19.** *Continuous method: cracking pattern of a restrained beam under early age shrinkage [SEL 14]*

In addition to these two approaches, special attention must be paid to consider the uncertainties associated with the prediction of crack opening and spacing. Reliability tools have been developed to be coupled with mechanical models. The difficulties associated with model coupling and computation run times are key points which require improvement for the future use of these tools on industrial applications.

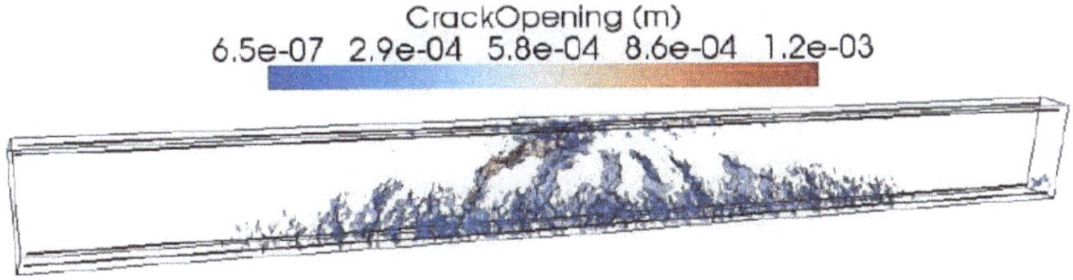


**Figure 1.20.** *Cracking pattern on a 3-point bending R.C. beam resulting from the use of a mixed method [OLI 15]*

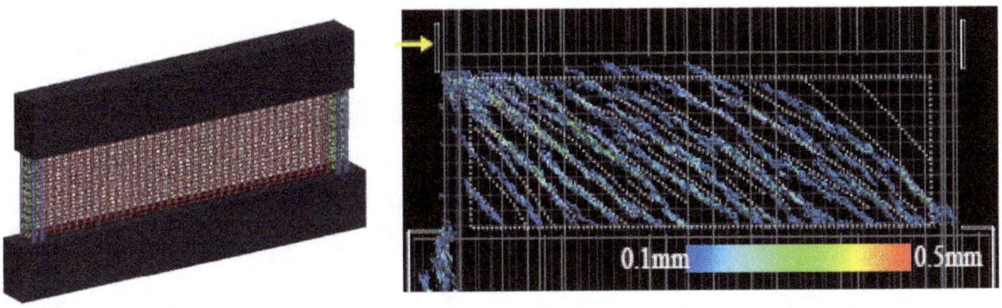


**Figure 1.21.** *Cracking pattern for SHW3 test using discrete elements method [YAM 14]*

## 1.3.2. *Benchmarks and workshops*

Following international benchmarking, a first workshop on the control of cracking in reinforced concrete structure – ConCrack1 – was held in December 2009. This promoted new relationships and increased exchange of information with the international scientific community. The performance of existing models was evaluated and compared to test results from the existing experiments. During this workshop, researchers presented the available modeling tools and exchanged their know-how. As a conclusion of the workshop, it was decided, based on the experiments derived from the CEOS.fr project, to organize an international benchmark exercise with the modeling of the behavior of the test mock-ups and large test bodies. The following were considered for the benchmark test: RL1 – large beam loaded in flexion with free shrinkage, RG8 – large beam with restrained shrinkage, and SHW3 – 1/3 scale shear wall under cyclic loading. A restitution workshop – ConCrack2 – was held in June 2011 and the main results were published in a special issue of the EJECE journal (September 2014).

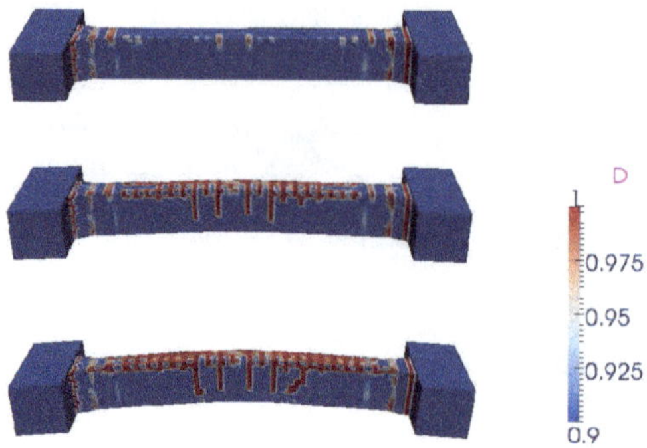


**Figure 1.22.** *Cracking evolution upon bending after restrained shrinkage operated on the beam RG8 (ConCrack2, Schreffler, Sciumé, Pesavento University of Padova)*

On the same theme, a further French-Japanese workshop – ConCrack 3 – was co-organized on THM effects in March 2012.

In March 2014, a final workshop – ConCrack 4 – co-sponsored by the European Joint Research Centre was held in Ispra (Italy). The main results obtained during the CEOS.fr program and related civil works in Europe were presented. Based on the results presented, standard rules for reinforced concrete, as applied to special structures, were proposed to a panel of international experts. In conclusion, it was decided to issue "Guidelines for the control of cracking in reinforced concrete structures".

### 1.3.3. *Numerical experiments*

The numerical simulations performed showed that the models are able to adequately represent the behavior of the test specimens (static, cyclic and thermo-hydro-mechanical loads). It was decided to use these models to extend the experimental program through numerical experimental models.

A numerical simulation program was planned to supplement the results obtained from physical experiments, which analyze the effects of various parameters on cracking phenomena.

In particular, the effects of geometry, concrete cover, reinforcement ratio, concrete strength, etc. were studied.

Figure 1.23 gives an example of the impact of two different reinforcement ratios on crack occurrence under restrained shrinkage.

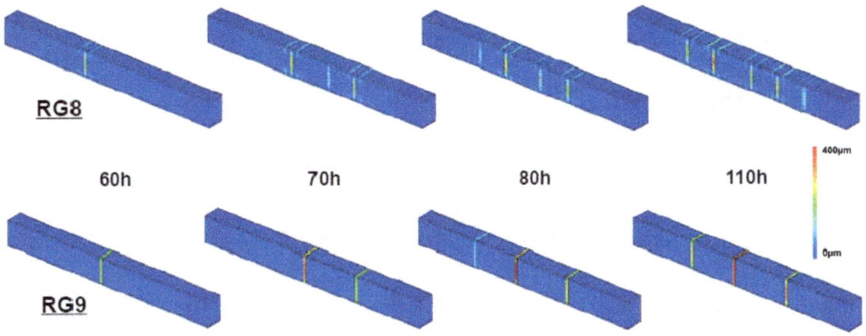


**Figure 1.23.** *Coupled thermo-hydro-mechanical modeling to forecast the onset of cracking at early age under the effect of a restrained shrinkage for two beams to different ratios of reinforcement [SEL 14]*

## 1.4. Engineering

In addition to the above data from tests on large bodies and modeling simulation, other results and data were used to support the CEOS.fr project, including:

– tests performed in laboratories (EPFL tests, Durham University tests) were used for the comparison with the results given by EC2 and MC2010 formulae dedicated to crack width and crack spacing calculation;

– for long-term results related to drying shrinkage and creep, a review of in-service measurement data from the monitoring and visual checks of existing structures has been used to supplement the above CEOS.fr test results. This data has been sourced from feedback from in-service structures: cooling towers, nuclear power plants (NPP), containments and associated large-scale models, such as the MAEVA model for NPP inner containment, concrete nuclear waste containers, cantilever prestressed decks and bridge piers.

Through the use of CEOS.fr project test and simulation results, complemented by test laboratory results and other inputs from the existing structures, the participants of the CEOS.fr project have formulated engineering proposals to supplement the EC2 and MC2010 formulae. These proposals are presented in the following chapters.

## 1.5. Database and specimen storage

### 1.5.1. *Database CHEOPS*

Data related to the CEOS.fr project have been captured on an experimental and numerical database, CHEOPS, developed by the Numerical Engineering & Consulting Society, NECS (necs.fr).

The data are selected for the use of numerical benchmarking. They are presented in the form of experimental identification data sheets (FIDEX). Each data sheet provides the following information:

– summary;

– geometrical and technical characteristics (formwork setup and concrete reinforcement);

– material properties (test results and value data sheet);

– experimental procedures (test bench, environmental conditions and photos);

– measurements (description, sensors, patterns, numerical files, comments and analysis).

As a registered user, the database can be publicly accessed at the following website: https://cheops.necs.fr/

### 1.5.2. *Specimen storage (Renardières site)*

Following testing, the RL and RG blocks have been stored by EDF on the Renardières Site located near Fontainebleau, France.

2

# Hydration Effects of Concrete at an Early Age and the Scale Effect

## 2.1. Hydration effects of concrete at an early age

Mechanical effects due to cement hydration at an early age depend on the limit conditions that structural elements are subjected to. Where shrinkage is restrained during the cooling phase, tensile stresses are generated and may lead to cracking (see Chapter 8). Where shrinkage is unrestrained, temperature differentials should be considered.

Three temperature differentials are identified and are to be taken into account under THM effects at an early age, resulting from cement hydration in concrete:

– the temperature difference between the core and the surface, mainly during and after the heating phase, or after the formwork removal or during the cure;

– the difference between the mean temperature of a new concrete lift and the temperature of the previous layer which supports the new lift; this difference occurs mainly 10 to 30 days after pouring depending on the concrete element thickness; and

– the mean temperature difference between two concrete elements, which have a different thickness, and are poured at the same time.

The three temperature differences can be estimated with reasonable agreement, or using graphs, nomograms or by simple thermal calculations, taking into account the hydration properties of the cement and the cement ratio.

NOTE 2.1.–It should be verified that an excessive temperature does not lead to the formation of chemical deleterious compounds due to the delayed ettringite formation (DEF). The model code MC2010 points out the phenomena in MC2010 section 7.8.9 "Delayed Ettringite Formation" (DEF):

> "If the concrete is exposed to high temperature over a certain period in its early life, delayed ettringite formation can occur depending on humidity and concrete mix design (alkali content, chemical composition of cement, use of additions, etc.)."

For design, the Provisional Recommendations on prevention of silica alkali reaction disorders can be also referred to [LCP 07].

### 2.1.1. *Global heating and cooling of a concrete element*

#### 2.1.1.1. *($T_{max}$ – $T_{ini}$) determination*

The temperature rise at a given point, which is the differential between the maximum temperature $T_{\max}$ and the initial temperature $T_{\text{ini}}$, can be determined according to the following step-by-step approach depending on the accuracy achieved, the possibility and the difficulty of the design calculations and their cost:

– a simplified approach (no effect due to formwork isolation or formwork removal);

– a one-dimensional finite difference calculation (allowing to consider exchange coefficients on facing); and

– a finite element calculation where the details of the geometry of the considered element are required.

In the case of a simplified approach, the method described in LCPC-IFSTTAR [LCP 07] "Recommendations on the silica alkali reaction prevention" may be used. For this method, the details of the geometry of the considered element, the mix concrete design, and the cement properties are required. An example of application of this method is presented in Box 2.1.

The IFSTTAR application for Android tablets and smartphones may also be used for the simplified approach.

Two more precise methods are not described in these Guidelines. They consist in assessing the maximum temperature rise resulting from cement hydration, by a one-dimensional finite difference (method 2) or a finite element calculation (method 3).

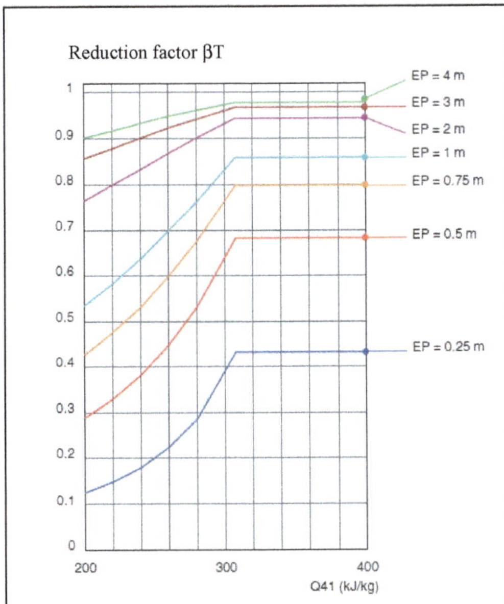


For a simplified approach, the reduction factor $\beta_T$ may be estimated according to the cement content using graphs such as the graph contained in the technical guide published by IFSTTAR-LCPC [LCP 07], and presented here. For example

- thickness = EP = 0.50 m
- cement content: 400 kg/m$^3$

$T_{adiab} = 71°C$

$T_{initial} = 18°C$

$Q\infty/\ \rho_C = 53°C$

$Q\infty = 332$ kJ/kg

$Q\infty \approx Q41$

$\beta_T = 0.7$

$T_{max} - T_{init} = 0.7 \times 53°C \approx 37°C$

**Box 2.1.** *Estimation of reduction factor*

## 2.1.2. *Differential temperature between concrete core and surface*

Cracking may occur few days after concreting as a result of temperature difference as shown by the stress profile. Cracking depends on the insulation quality of the formwork and on the cure duration (Figure 2.1).

In the short-term (around three days), the surface is cooler than the core of the concrete. Uncontrolled cracking on the surface may occur without concrete reinforcement, with an overly quick formwork removal or a cure stop which is too early (see Figure 2.1 (a)). In this case, the area in tension $A_c$ is limited to approximately 20% of the thickness of the element on each face.

In the long-term, the temperature of the core decreases down to the ambient temperature. This cooling takes 10 to 30 days according to the element thickness. Generally, a lift or a raft restrains the element. A slab may also be end-restrained by walls located at both ends. This restraint may result in internal cracks at the core of the element. By the time, these cracks may expand to the surface according to the restraint value (see Figure 2.1(b)).

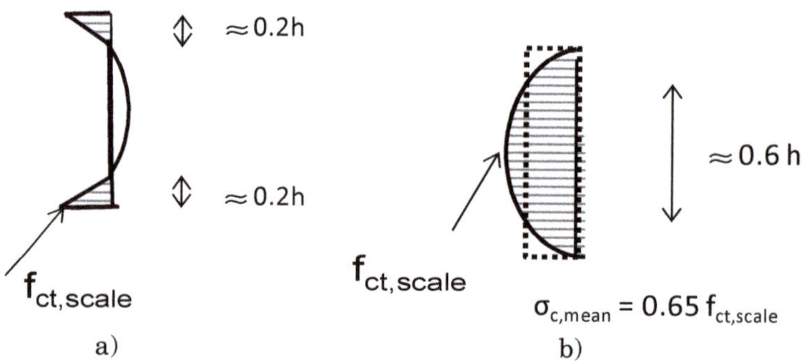


**Figure 2.1.** *Gradient between core and surface: a) immediately after formwork removal; b) during the phase reaching the restraint equilibrium*

The cooling rate may be estimated with a thermal calculation that simulates hydration then formwork removal. This calculation may be three-dimensional or two-dimensional. But for the current cases, a one-dimensional calculation can be made with an Excel spreadsheet to simulate the temperature rise of a wall during hydration then cooling.

For example, assuming that the cement content is 400 kg/m$^3$, the hydration heat is $Q_\infty$ = 332 kJ/kg, the concrete temperature is 15°C, the constant external temperature is 15°C, the formwork removal is carried out after 7 days (the exchange coefficient is 4 W/m/°C before the form removal and 8W/m/°C after removal), and the cooling rate is:

– 7 days for a wall 0.40 m thick;

– 14 days for a wall 0.75 m thick;

– 21 days for a wall 1 m thick;

– 28 days for a wall 2 m thick.

For a 1 m thick wall, the following graph shows the change of the maximum temperature, the mean temperature and the surface temperature (the inside and outside temperatures are the same and the internal and external surface temperatures are the same).

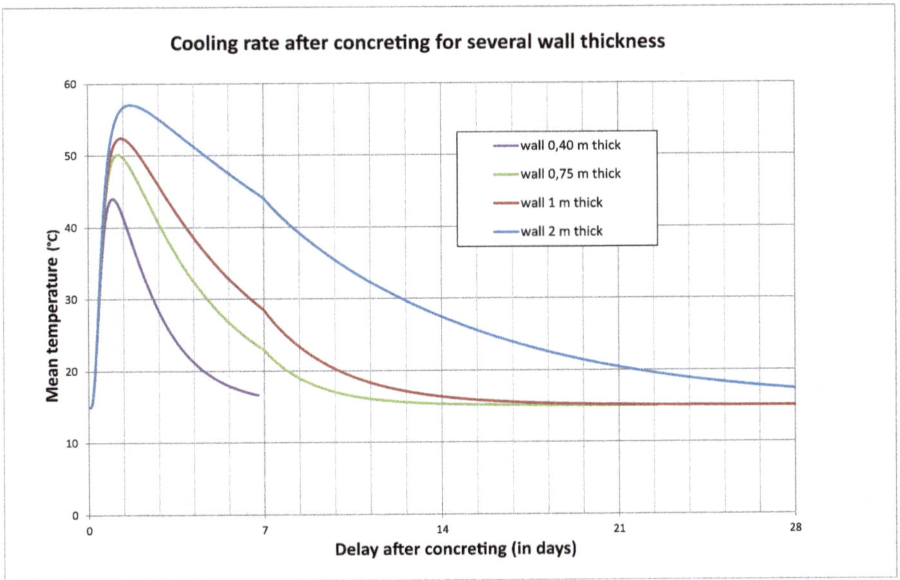


**Figure 2.2.** *Cooling after concreting for various wall thicknesses*

Another source of the gradient between facing and concrete core is due to the daily variation of outside temperature. This variation would affect a depth of 300 mm from the surface of a slab or a wall in the case of a relatively humid concrete with diffusivity $a=\lambda$ / $\rho c=92010^{-9}$ $m^2/s$ derived from the following thermal characteristics: $\lambda = 2.3$ W/m°K; c = 1000 J/kg; $\rho = 2500$ kg/m$^3$.

NOTE 2.2.– For the combination of mechanical actions and imposed deformations, see Chapter 9 “Effects of various phenomena combination”.

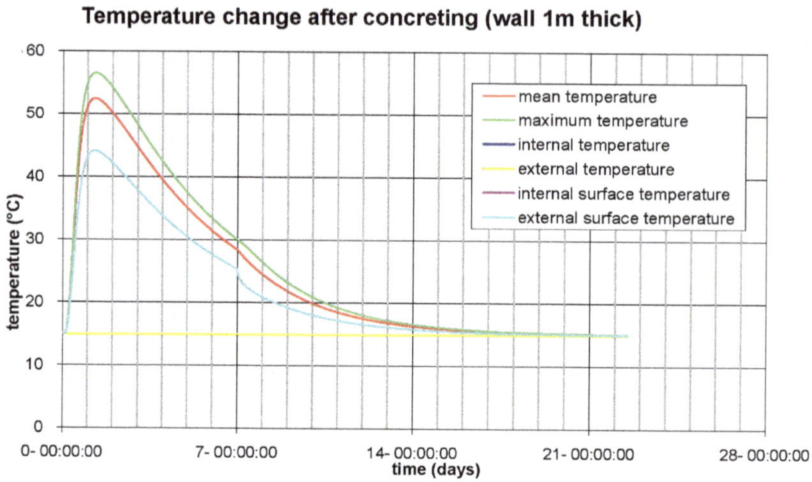


**Figure 2.3.** *Temperature change after concreting*

## 2.2. Scale effect

### 2.2.1. *Scale effect principle*

The heterogeneity of concrete and its non-linear post-peak behavior beyond its tensile limit strength lead to a decrease of crack resistance when the volume under tensile loading is large. The first crack is generally linked to the concrete nominal tensile strength, measured from tests on specimens sometimes of standard size, but usually of small volumes. Tensile strength measured on concrete massive elements (i.e. first crack) compared to the experimental value of tensile stress obtained from small specimens is generally decreasing with the increasing of the concrete volume.

This phenomenon known as the Weibull scale effect is based on the theory of the weakest link [CAR 95, BAZ 04, BAZ 05, SEL 13, SEL 14]. It had been identified and quantified in particular by Rossi [ROS 94] from axial tensile tests performed on cylinders of several sizes and made of various concrete formulations. According to size and concrete formulation, 7 to 12 samples were tested. The test results highlighted the following relationships:

– decreasing of the mean tensile strength of the concrete ($f_{ctm}$) in accordance with the considered volume of the element;

– decreasing of the standard deviation of the experimental tensile stress while the concrete volume increases, for the same number of samples;

– only the tensile stress appears to be affected by the scale effect; on one hand, the elastic modulus does not seem directly affected; on the other hand, the coefficient of dispersion of the modulus decreases as the volume and the compressive strength of the concrete increases; and

– the lower the concrete compressive stress ($f_{cm}$), the higher the scale effect.

As a result from these findings, the scale effect is concrete quality dependent: the more homogeneous the concrete, the lesser the reduction of tensile strength according to the volume.

While Rossi linked the concrete heterogeneity to the coarsest grain size, [VAN 00] showed that scale effect is minimized for dry samples: porosity is then probably a key factor as drying induces micro cracking.

Occurrence of the first crack highlights the scale effect for massive blocks with a large concrete volume subjected to tensile stress. Thus, the probability of encountering a low concrete tensile strength is high (weakest link effect). But when the stabilized cracking stage is reached (see Figure 4.1, point S), the volume of un-cracked concrete between two cracks decreases and the probability to meet low tensile strength decreases also. Then the concrete tensile strength is close to the tensile strength measured on the laboratory test specimen.

### 2.2.2. *Calculating scale effect according to Weibull theory*

#### 2.2.2.1. *Weibull theory*

Consideration of a random spatial field of mechanical characteristics for an element under a stress field is a way to explain the variability of the limit tensile stress leading to the first crack in a large volume element [ROS 12].

The mean limit stress obtained from several random field implementations decreases significantly with the element size, as the probability to meet a weak part in the zone under the tensile stress gradient is higher. However, referring to random spatial fields is not realistic for an analytical calculation. Then Weibull theory could be considered.

Weibull theory is a simplified theory, which assumes that the local concrete damage generates the first crack; indeed the different parts of the concrete massive element could not give each other mutual assistance. This hypothesis shows the limit of the theory. However, this theory has been useful to develop research on concrete related to this topic since 1973 [LHE 73].

The scale effect was clearly identified on micro-concrete tests by [ROS 94] for concrete compressive strengths ranging from 30 to 130 MPa. Simple laws of the following type have been derived from these tests relating to tensile strength:

$$\left[\frac{f_{ctm}^{veq}}{f_{ctm}^{vref}}\right] = \frac{V_{ref}}{V_{eq}} \qquad [2.1]$$

where

– $f_{ctm}$ is the mean concrete tensile strength in MPa for the considered volume $V_{ref}$ or $V_{eq}$;

– $f_{cm}$ is the mean concrete compressive strength given that $f_{cm} = f_{ck} + 8$ MPa according to EC2 (see section 2.2.3);

– $V_{ref}$ is the reference volume of tests (for example volume of specimen);

– $V_{eq}$ is the equivalent volume taken into account for the considered concrete element (see hereafter its calculation);

– k is the Weibull exponent.

$$k = (0.25 - 3.6.10^{-3} f_{cm} + 1.3.10^{-5} {f_{cm}}^2)^{-1}$$

This formula may be applied to concrete of class resistance equal to 25 MPa with a reasonable agreement when test results are missing.

#### 2.2.2.2. *Volume $V_{eq}$ to be considered*

Rossi considered specimens of small sizes. He assessed that the volume of the coarsest grain was relevant geometrical data. However, the scale effect at these low scales may be generally scaled up, by referring to a reference volume ($V_{ref}$) close to the specimen volume ($16 \times 32$ $cm^3$).

For large-size elements, such as CEOS.fr test blocks, the concrete tensile strength does not decrease to zero when the element size expends indefinitely, contrary to the prediction of Weibull theory.

An available solution for engineering consists of introducing the concept of "maximum volume" [SEL 12]:

> *The maximum volume to be considered in a Weibull law is around 2 $m^3$.*

This assumption enables retrieval of the values of tensile strength measured on large specimens and ties (Figure 2.4).

Figure 2.4 plots the limit tensile stress corresponding to the tensile failure of tested specimens (specimens performed by Rossi) and to the initiation of the first crack in large-sized tested structures (Milevaz ties 1 m × 0.42 m × 5 m). This limit tensile stress is compared to theoretical diagrams showing the evolution of the tensile strength $f_{ctm}$ according to the scale effect (in Figure 2.4, dotted red and blue curves). The figure, drawn on the base of experimental results and Rossi's formula, highlights the significant decrease of $f_{ctm}$ values for two concretes of mean compressive resistance 35 and 55.8 MPa. The evolution is representative of the scale effect phenomena, with a clear decrease of the tensile strength when the size of the concrete element increases.

#### 2.2.2.3. *Dispersion around the mean value with scale effect*

For the current elements, scale effect is implicitly taken into account in EC2 and MC2010, notably concerning the characteristic value $f_{ctk,0.05}$ which ensures that no more than 5% of founded values are less than $f_{ctk,0.05}$.

For concrete elements of large volume (massive components), the probability to observe the occurrence of a first crack at a value $f_{ct,inf}$ less than $f_{ctk,0.05}$ is higher because the corresponding concrete volume is larger and then the scale effect is more pronounced.

Figure 2.4 moreover shows the decrease of standard deviation with the volume.

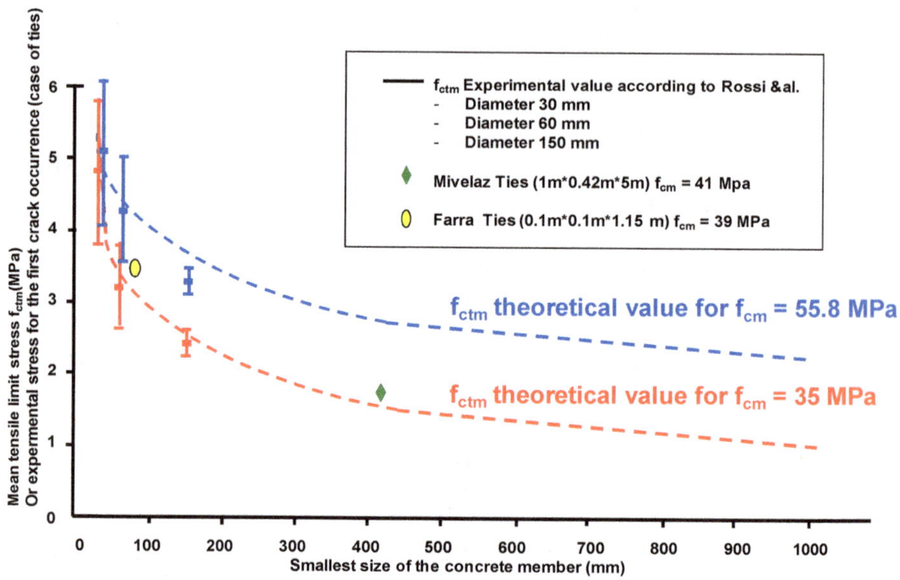


**Figure 2.4.** *Evolution of mean tensile strength ($f_{cm}$) according to the size of the element evaluated by its smallest dimension. For tests performed by Rossi, the standard deviation obtained from seven tests is plotted*

For small size elements, Rossi proposed a formula for assessing the dispersion, the ratio between standard deviation and the mean value [ROS 94].

$$B = 4.5 \cdot 10^{-2} + 4.5 \cdot 10^{-3} f_{cm} - 1.8 \cdot 10^{-5} f_{cm}^2 \qquad [2.2]$$

where:

– $f_{cm}$ is the mean value of the compressive concrete strength;

– B is the dispersion.

(See section 3.6, "Example of application of cracking formulae for a tie under tension").

For large concrete volumes (massive elements), as experimental standard deviation decreases with the tensile strength according to the scale effect, the following relation may be applied with a reasonable agreement to a standardized specimen:

$$\text{coefficient of dispersion} = \text{standard deviation}/f_{ctm} = \text{constant} \qquad [2.3]$$

This means that the standard deviation is proportional to the mean tensile stress $f_{ctm}$.

NOTE 2.3.–

– According to Weibull theory, the Weibull's coefficient, k is a constant regardless of the scale, allowing [2.3] to be used; and

– assuming that the coefficient of variation is 0.2 for a normal distribution, the relations giving the fractiles 5% and 95%, namely $f_{ctk,0.05}$ and $f_{ctk,0.95}$ are the same as those provided by EC2. The coefficient of variation is in accordance with experimental values measured by Rossi with the smallest specimens – but not with specimens of larger diameter (150 mm), for which this coefficient is lower.

#### 2.2.2.4. *In summary*

For assessing the influence of the scale effect on the mean value of the concrete tensile strength, the value $f_{ctm,scale} = f_{ctm}^{Veq}$ should be calculated by applying the above methodology based on [2.1]. This value replaces the mean value $f_{ctm}$ in the equations given by EC2 and MC2010, particularly for calculating the crack width.

To assess the first crack occurrence, in addition a sufficient estimation of the standard deviation is required in order to evaluate the fractile $f_{ct0.05}^{Veq}$, either preferably by referring to an adequate number of tensile tests for a given concrete, or either, failing sufficient data, by using the above equations to assess the dispersion ([2.2] or [2.3] according to the size of the concrete element).

### 2.2.3. *Worked examples of calculation with the scale effect according to the Weibull model*

In the case of a massive concrete element of the same dimensions as an RL beam of the CEOS.fr project, the purposes of these examples are:

– to calculate the mean tensile strength resulting from the scale effect; and

– to evaluate the strength corresponding to the first crack assessed according to the fractile 5% for different concrete strength classes defined by their characteristic strength.

### 2.2.3.1. *Data*

– Dimensions of RL beam:

l = 6.00 m

b = 1.60 m

h = 0.80 m.

To quantify the consequences of uncertainty concerning the height $h_{cef}$, two calculations are carried out:

1st calculation $h_{cef} = 30$ cm

2nd calculation $h_{cef} = 40$ cm.

– Initial values of tensile strengths are given in Table 2.1 of EC2 according to the concrete strength class specified by their characteristic strength value $f_{ck}$; the other strengths are derived from $f_{ck}$ in accordance with EC2 as follows:

| | | |
|---|---|---|
| $f_{cm}$ | $= f_{ck} + 8$ MPa | Mean compressive strength |
| $f_{ctm}$ | $= 0.30 f_{ck}^{(2/3)}$ if ≤ C50/60 | Mean tensile strength |
| | $= 2.12.\ln(1+f_{cm}/10)$ if > C50/60 | |
| $f_{ctk;\,0.05}$ | $= 0.7\, f_{ctm}$ | fractile 5% |
| $f_{tck;0.95}$ | $= 1.3\, f_{ctm}$ | fractile 95% |

**Table 2.1.** *Tensile strength values according to characteristic strength values*

– It is assumed that the values used in calculating $f_{ck}$ result from split tests performed on specimens, so the reference volume is reduced to $V_{eq} = 300\ cm^3$.

Table 2.1 of EC2 provides the values of $f_{ctm}$ in order to assess the influence of concrete strength classes on scale effect.

For a given structure, the strength class is generally specified and studied in the laboratory so that the mean value $f_{ctm}$ is obtained from tensile tests on specimens at 3, 7 or 28 days.

Similarly, tests performed in a research laboratory rely either on the mean value of axial tensile tests, $f_{ctm}$, or on the limit tensile value from split tests. Relationships between axial and split test values are proposed; the adopted value is obtained from tests performed at 28 days:

$f_{tsp} = 0.387 \cdot f_{cm}^{0.63}$ [ARI 06]

$f_{ctm} = 0.363 f_{cm}^{0.527}$ [FAR 95]

Calculating $f_{ctm}^{Veq}$:

$f_{ctm}^{Veq} = f_{ctm}^{Vref} \left(V_{ref}/V_{eq}\right)^{1/k}$

$f_{ctm}^{Vref} =$ tensile stress measured on $V_{ref}$

$f_{ctm}^{Veq} =$ tensile stress taking account of scale effect

$V_{ref} = 300\ cm^3$, in the case of a split test on concrete cylinders, with 11 cm diameter and 22 cm hight; indeed, for a split test only a part of the specimen is under direct tension, as shown by the stress field at the failure time [SEL 12] (see the blue zone aligned with the upper apex line of the cylinder in Figure 2.5).

As mentioned for elements of large dimensions, the volume $V_{eq}$ is restrained to 1.25 m for each dimension, given that only the volume under tension is considered [SEL 12]. In the case of a mechanical loading, the effective height, $h_{c,eff}$, is concerned. When no complete finite element calculation is available, the volume $V_{eq}$ may be limited as a first approximation, to 1.25 m for each dimension and if necessary, modified in accordance with the shape of stress diagram in the zone in tension:

$$V_{eq} = r.\min(1.25;l).\min(1.25;b).\min(1.25;h_{cef}).$$

where:

– $l$, $b$, and $h_{c,eff}$ are respectively the length, width and effective height of the volume in tension;

– $r$ is a coefficient of reduction taking account of the non-homogeneous stress field before the first crack;

- $r = 1$ for axial tension,

- $r = 1/(1+k)$ in the case of pure bending, k being the Weibull exponent;

– $r$ may be calculated by linear interpolation between these two values in the case of combined bending. The value of r is demonstrated in section 2.2.4 of this chapter.

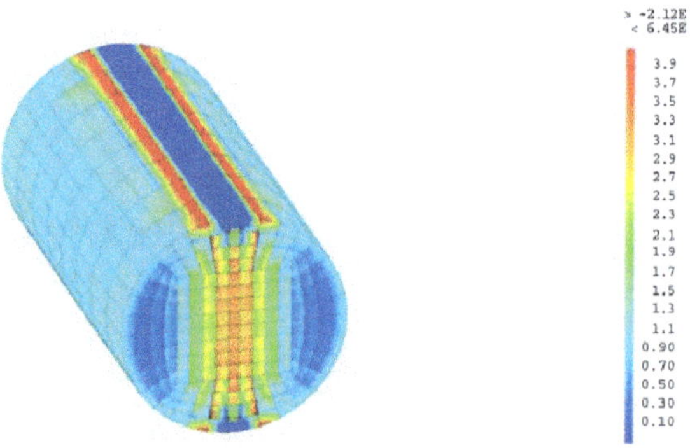


**Figure 2.5.** *Volume in tension during a split test (red zone along the cylinder)*

NOTE 2.4.–When the risk of cracking is predominant at an early age and is due only to THM effects, the volume $V_{eq}$ to be considered is the volume which results from the temperature gradient during the cooling phase significant enough to generate cracking. This risk depends on the phases of formwork removal.

Two cases are presented in the following paragraph:

– the case of bending: the first cracking is supposed to result from a flexural loading in the tensile one, $h_{cef}$ high; and

– the case of axial tension under THM effects at the core: resulting from differential cooling during formwork removal, the concrete settling at the core under the temperature creates a zone in tension, $h_{cef}$ high.

*Calculating* $f_{ct\,0.05}^{Veq}$

Assuming that the coefficient of dispersion of the normal distribution of $f_{ctm}$ is constant when the equivalent volume $V_{eq}$ increases:

$$\frac{s^{Vref}}{f_{ctm}^{Vref}} \cong \frac{s^{Veq}}{f_{ctm}^{Veq}}$$

Where *S* is the standard deviation.

Then

$$s^{Veq} = s^{Vref} \cdot \frac{f_{ctm}^{Veq}}{f_{ctm}^{Vref}}$$

Assuming that the distribution of $f_{ctm}$ is normal,

$$f_{ctk0.05} = f_{ctm} - 1.65\ S_{Vref}\ \text{(fractile 5\%)}$$

and

$$s^{Vref} = \frac{0.3}{1.65} \cdot f_{ctm}^{Vref}$$

$$s^{Veq} = s^{Vref} \cdot \frac{f_{ctm}^{Veq}}{f_{ctm}^{Vref}} = \frac{0.3}{1.65} \cdot f_{ctm}^{Vref} \cdot \frac{f_{ctm}^{Veq}}{f_{ctm}^{Vref}} = \frac{0.3}{1.65} \cdot f_{ctm}^{Veq}$$

$$f_{ct0.05}^{Veq} = f_{ctm}^{Veq} - 1.65 \cdot s^{Veq} = 0.7 \cdot f_{ctm}^{Veq}$$

This relation is the relation given in Table 2.1 of EC2 (as already mentioned in section 2.2.3.2).

$$f_{ct,0,05} = 0.7\ f_{ctm}$$

It is noted that:

– the coefficient of deviation is supposed to be constant, for example [2.3] is valid;

– the relation $f_{ctk,0.05} = 0.7\ f_{ctm}$ is sufficiently representative of reality to assess the influence of concrete compressive class on scale effect; for a project designed with a given concrete compressive class, it is better to refer to $f_{ctm}^{Veq}$ measured on specimens tested in a laboratory.

APPLICATION.–

1) Case where $h_{ceff}$ = 30 cm.

For a RL beam tested in the CEOS project:

l = 6.00 m

b = 1.60 m

h = 0.80 m

$h_{ceff}$ = 0.30 m

$V_{eq}$= r.min (1.25;6.00).min (1.25;1.6).min(1.25;0.30)

$V_{eq}$ = r.1.25 x 1.25 x 0.30 = 0.468 $m^3$ = 468.$10^3$ cm3

$V_{ref}/V_{eq}$ = (0.641.$10^{-3}$)/r

| Stress (MPa) | *Case 1: $h_{ceff}$=30cm* | | | | |
|---|---|---|---|---|---|
| **$f_{ck}$** | **30** | **40** | **50** | **60** | **80** |
| $f_{cm}$ | 38 | 48 | 58 | 68 | 88 |
| **$f_{ctm}$** | **2.9** | **3.5** | **4.1** | **4.4** | **4.8** |
| $1/k$ | 0.1320 | 0.1072 | 0.0849 | 0.0653 | 0.0339 |
| $k$ | 7.6 | 9.3 | 11.8 | 15.3 | 29.5 |
| Coefficient r in bending | 0.1166 | 0.0968 | 0.0783 | 0.0613 | 0.0328 |
| Coefficient r in tension | 1 | 1 | 1 | 1 | 1 |
| $V_{ref}/V_{eq}$ tension | 0.00064 | 0.00064 | 0.00064 | 0.00064 | 0.00064 |
| $V_{ref}/V_{eq}$ bending | 0.00550 | 0.00662 | 0.00819 | 0.01046 | 0.01957 |
| $(V_{ref}/V_{eq}$ tension$)^{1/k}$ | 0.37896 | 0.45483 | 0.53555 | 0.61866 | 0.77955 |
| (Vref/Veq bending)$^{1/k}$ | 0.50324 | 0.58415 | 0.66491 | 0.74240 | 0.87524 |
| **$f_{ctm}^{Veq}$ (tension)** | **1.1 (0.4 $f_{ctm}$)** | **1.6 (0.45 $f_{ctm}$)** | **2.2 (0.5 $f_{ctm}$)** | **2.7 (0.6 $f_{ctm}$)** | **3.7 (0.8 $f_{ctm}$)** |
| **$f_{ctm}^{Veq}$ (bending)** | **1.5** | **2.0** | **2.7** | **3.3** | **4.2** |
| **$f_{ct0.05}$ (tension)** | **0.8** | **1.1** | **1.5** | **1.9** | **2.6** |
| **$f_{ct0.05}$ (bending)** | **1.1** | **1.4** | **1.9** | **2.3** | **2.9** |

**Table 2.2.** *Calculation of scale effect for a beam 0.30 m high*

*Calculation of k exponent of Weibull equation*

According to the relation established by Rossi:

$$k = (0.25 - 3.6.10^{-3} f_{cm} + 1.3.10^{-5} f_{cm}{}^{2})^{-1}$$

The following values are obtained for different values of concrete characteristic strengths to which EC2 associates a mean value of the tensile strength $f_{ctm}$.

2) Case where $h_{ceff}$ = 40 cm.

l = 6.00 m

b = 1.60 m

h = 0.80 m

$h_{ceff}$ = 0.40 m

$V_{eq}$= r.min (1.25;6.00).min (1.25;1.6).min(1.25;0.40)

$V_{eq}$ = r.1.25 × 1.25 x 0.40 = r.0.625m$^3$ = r.625.10$^3$ cm3

$V_{ref}/V_{eq}$ = (0.480.10$^{-3}$)/r

| Stress (MPa) | *Case 2: $h_{ceff}$ = 40cm* | | | | |
|---|---|---|---|---|---|
| **$f_{ck}$** | **30** | **40** | **50** | **60** | **80** |
| $f_{cm}$ | 38 | 48 | 58 | 68 | 88 |
| **$f_{ctm}$** | **2.9** | **3.5** | **4.1** | **4.4** | **4.8** |
| *1/k* | 0.1320 | 0.1072 | 0.0849 | 0.0653 | 0.0339 |
| *K* | 7.6 | 9.3 | 11.8 | 15.3 | 29.5 |
| Coefficient r in bending | 0.1166 | 0.0968 | 0.0783 | 00613 | 0.0328 |
| Coefficient r in tension | 1 | 1 | 1 | 1 | 1 |
| $V_{ref}/V_{eq}$ tension | 0.00048 | 0.00048 | 0.00048 | 0.00048 | 0.00048 |
| $V_{ref}/V_{eq}$ bending | 0.00412 | 0.00497 | 0.00614 | 0.00784 | 0.01467 |
| $(V_{ref}/V_{eq}$ tension$)^{1/k}$ | 0.36484 | 0.44102 | 0.52262 | 0.60714 | 0.77199 |
| $(V_{ref}/V_{eq}$ bending$)^{1/k}$ | 0.48449 | 0.56642 | 0.64886 | 0.72858 | 0.86676 |
| **$f_{ctm}^{V_{eq}}$ (tension)** | **1.1** | **1.5** | **2.1** | **2.7** | **3.7** |
| **$f_{ctm}^{V_{eq}}$ (bending)** | **1.4** | **2.0** | **2.7** | **3.2** | **4.2** |
| **$f_{ct.0.05}$ (tension)** | **0.8** | **1.1** | **1.5** | **1.9** | **2.6** |
| **$f_{ct.0.05}$ (bending)** | **1** | **1.4** | **1.9** | **2.2** | **2.9** |

**Table 2.3.** *Calculation of scale effect for a beam 0.40 m high*

NOTE 2.5.–

– The reduction factor of the scale effect is greater for low concrete strength classes; for $h_{ceff}$ = 0.30 m, the reduction factor is 0.4 for $f_{ck}$ = 30 MPa and only 0.8 for $f_{ck}$ = 80 MPa (see also Table 6.2 for other values for the reduction factor);

– the results are relatively insensitive to a variation of the equivalent volume V: in this case, 0.625 $m^3$ instead of 0.468 $m^3$, that is +25% with respect to 0.625 $m^3$; and

– to be in bending instead of to be in tension influences significantly the results; for example, for an element in bending presenting a zone in high tension, narrower than an element in tension, its first cracking strength is statistically greater.

### 2.2.4. *Application of Weibull integral in bending and in tension*

The mathematical definition of the equivalent volume is taken from the logarithmic integral of probability of survival (Ps = 1- Pt):

$$\ln(Ps)\, V_{eq} = \int_V \ln(Ps)\, dV$$

with a Weibull distribution $Ps = \exp\left(\frac{\sigma}{f_{ctm}(vref)}\right)^k$

$$V_{eq} = \frac{\int_V \left(\frac{\sigma}{f_{ctm}^{Vref}}\right)^k dV}{\left(\frac{\sigma_{max}}{f_{ctm}^{Vref}}\right)^k}$$

In this definition, $\sigma_{max}$ is the maximum stress inside of the calculated element, and $\sigma$ is the stress field.

– In axial tension, $V_{eq}$ = V, excluding the maximum volume; but the maximum volume should be applied for large volumes that are involved [SEL 12],

– in bending, the diagram is linearly dependent of the height h; it is assumed for example that $\sigma = \sigma_{max}\frac{y}{h}$, $h$ being the height in tension.

By introducing this mathematical expression in the above $V_{eq}$ equation, we obtain:

$$V_{eq} = \frac{\left(\frac{\sigma max}{f_{ctm}^{Vref}}\right)^k \iint_{y=0}^{h} \left(\frac{y}{h}\right)^k dxdydz}{\left(\frac{\sigma max}{f_{ctm}^{Vref}}\right)^k} = \int_{b.lc} dxdz \int_{y=0}^{h} \left(\frac{y}{h}\right)^k dy$$

$$V_{eq} = \frac{1}{1+k} V$$

# 3

# Cracking of Ties

## 3.1. Design values and limit values

The concepts of mean, characteristic and maximum values in relation to the crack spacing and width are not clearly defined in the current standards and as a result this can cause problems for their users. It is proposed that the values used in the verification rules should be a "design value", an approach considered to be more in line with the intent of the standards. That is, $s_d$ for the crack spacing and $w_d$ for the crack width.

PROPOSAL 3.1.– Use the design values, $s_d$ and $w_d$, in the equations relating to cracking. The verification equation for the crack width is written as follows:

$$w_d \leq w_{lim}$$

## 3.2. Adjusting the design value for verification purposes

In both EC2 and MC2010, it would appear that the "design value" used for the crack width is in reality either a maximum characteristic value or one very close to such a value. This value is used for verification purposes in serviceability limit states and does not impact on the safety of the structure. In most cases, this verification is determined more by durability considerations or visual appearance. Hence, it follows that some form of crack treatment for a small number of cracks may be envisaged over the life of the works. Under these conditions, the design value may be based on a mean value or on a characteristic value with a greater probability of being exceeded. However, this approach is not desirable in works where water- or air-tightness requirements must be *justified* or *fulfilled* or *equivalent*.

## 3.3. Crack spacing equation

### 3.3.1. *Linear equation*

In the case of crack spacing, a linear equation containing two parameters, c (cover) and $\frac{\emptyset}{\rho_{s,eff}}$ (the ratio between the diameter of the reinforcement bars and the effective reinforcement ratio), of the type:

$$S_{r,max} = \widehat{K}_1 c + \widehat{K}_2 \frac{\emptyset}{\rho_{s,eff}} \qquad [3.1]$$

gives satisfactory results in comparison with experimental results obtained from a concrete tie with one reinforcement bar. The coefficients adopted in EC2 give values that are high by comparison with experimental results. While the equation given in MC2010 could be improved, it does give results that are closer to the experimental results.

This improved adequacy between the results of the equation in MC2010 and experimental results has also been found in the case of other test structures used in CEOS experiments.

PROPOSAL 3.2.– In the case of simple structural elements, an equation for the crack spacing consisting of a linear combination of c and $\frac{\emptyset}{\rho_{s,eff}}$ is simple and sufficient for assessing. However, the coefficients of this combination could be optimized but if not, the current equation given in MC2010 is considered to be acceptable.

$$S_{r,max} = 2c + \frac{1}{2} \cdot \frac{f_{ctm}}{\tau_{bm}} \cdot \frac{\emptyset_s}{\rho_{s,eff}} \qquad [3.2]$$

### 3.3.2. *Relationship between the maximum spacing $S_{r,max}$ and the mean spacing $S_{r,m}$*

The spacing between cracks is a random variable with a mean value $s_{rm}$ less than or equal to 2/3 of the maximum spacing $s_{r,max} = 2l_{s,max}$. The approximation $s_{rk,95} \approx s_{r,max}$ is therefore considered to be acceptable.

The ratio $s_{rk,95}/s_{rm}$ therefore lies in the range of 1.5 – 1.7. It depends on the coefficient of variation of concrete tensile strength. This observation provides further confirmation of the coefficient value of 1.7 suggested in EC2 and MC2010.

### 3.3.3. *Equation based on the MC2010 bond-slip relationship*

According to MC2010, the $\frac{f_{ctm}}{\tau_{bm}}$ ratio is assumed to be constant with a default value of $\frac{f_{ctm}}{\tau_{bm}} = \frac{1}{1.8} = 0.555$

However, MC2010 also proposes a bond-slip law in which the bond stress varies as a function of the relative slippage between the reinforcement bar and the concrete (Figure 3.1).

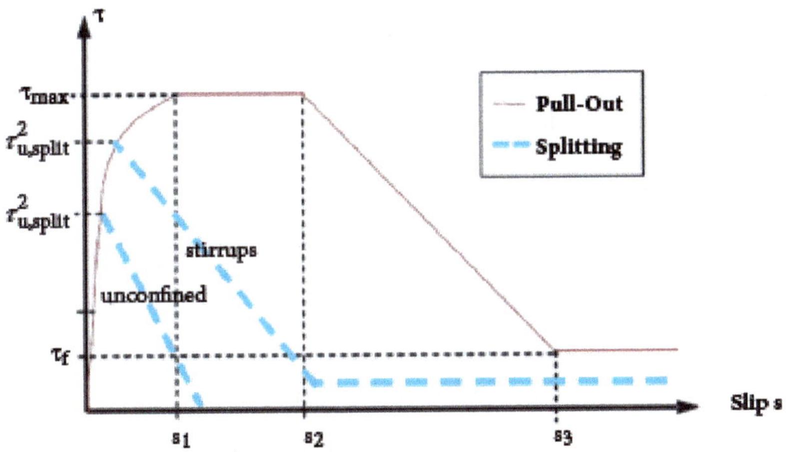


**Figure 3.1.** *Bond-slip relationship as given in MC2010 (Figure 6.1.1 of Model Code 2010).*

In simple cases, the value of the relative slippage between the reinforcement bar and the concrete $s = u_s - u_c$ corresponds to the first part of the curve in the above figure, and is given by the following expression:

$$\tau(s) = \tau_{max}\left(\frac{s}{s_1}\right)^{\alpha} \qquad \text{MC2010 Eq. [6.1.1.]}$$

If this bond-slip relationship is adopted, the maximum crack spacing is given by the following equation [BAL 93]:

$$s_{r,max} = 2l_{s,max} = 2\left(\frac{1+\alpha}{1-\alpha}\frac{\gamma f_{ctm}}{4\rho\tau_0}\right)^{\frac{1-\alpha}{1+\alpha}}(\phi g_0^{\alpha})^{\frac{1}{1+\alpha}}\left(\frac{2(1-\alpha)^2\left(1+\frac{\alpha_e\rho}{\gamma}\right)}{1+\alpha}\frac{\tau_0}{E_s}\right)^{\frac{-\alpha}{1+\alpha}} \qquad [3.3]$$

This equation is based on the use of $\rho = \frac{A_s}{A_c+A_s}$. However, if it is considered that the distribution of the stresses in the section of the concrete tie is not uniform, then it is necessary to consider the effective area $A_{c,eff}$ instead of $A_c$ or to introduce the ratio between the effective area and the total area $\gamma = \frac{A_{c,eff}}{A_c}$.

The effective reinforcement ratio is defined as follows:

$$\rho_{s,eff} = \frac{A_s}{A_{c,eff}+A_s} \approx \frac{A_s}{A_{c,eff}} = \frac{A_s}{A_c}\frac{A_c}{A_{c,eff}} \approx \frac{\rho}{\gamma} \qquad [3.4]$$

In the above equation for the spacing $s_{r,max}$, it is necessary to introduce $\rho_{s,eff}$ in place of $\rho$, which then gives:

$$s_{r,max} = 2l_{s,max} =$$

$$2\left(\frac{1+\alpha}{1-\alpha}\frac{\gamma f_{ctm}}{4\rho_{s,eff}\tau_0}\right)^{\frac{1-\alpha}{1+\alpha}}(\phi g_0^{\alpha})^{\frac{1}{1+\alpha}}\left(\frac{2(1-\alpha)^2\left(1+\frac{\alpha_e\rho_{s,eff}}{\gamma}\right)}{1+\alpha}\frac{\tau_0}{E_s}\right)^{\frac{-\alpha}{1+\alpha}} \qquad [3.5]$$

## 3.4. Equation for the mean differential strain

The equations given in the EC2 and MC2010 codes are as follows:

$$(\varepsilon_{sm} - \varepsilon_{cm})_{EC2} = \frac{\sigma_{s2} - k_t\frac{f_{ctm}}{\rho_{eff}}(1+\alpha_e\rho_{eff})}{E_s} \geq 0.6\frac{\sigma_s}{E_s} \qquad \text{EC2-1 Eq. [7-9]}$$

$$(\varepsilon_{sm} - \varepsilon_{cm} = \frac{\sigma_{s2} - \beta . \sigma_{sr}}{E_s}) \qquad \text{MC2010 Eq. [7.6-5]}$$

where:

$$\sigma_{sr} = \frac{f_{ctm}}{\rho_{s,eff}}(1 + \alpha_e \rho_{s,eff}) \qquad \text{MC2010 Eq. [7.6-6]}$$

which may be expanded to give:

$$(\varepsilon_{sm} - \varepsilon_{cm})_{MC2010} = \frac{\sigma_{s2} - \beta \frac{f_{ctm}}{\rho_{s,eff}}(1 + \alpha_e \rho_{s,eff})}{E_s}$$

Results obtained by the CEOS.fr project relevant to early age massive elements indicate that, when calculating the maximum crack width, the tension stiffening expression $\frac{\beta \sigma_{sr}}{E_s}$ needs to be modified by applying a weighting of 0.6 to the value of $\beta$ when the concrete is at early age. Equation [7.6.5] of MC2010, given in section 3.4, then becomes:

$$\varepsilon_{sm} - \varepsilon_{cm} = \frac{\sigma_{s2} - 0.6\beta . \sigma_{sr}}{E_s} \qquad [3.6]$$

PROPOSAL 3.3.– When determining the value of $f_{ctm}$ to be taken into account in the concrete tie or the part of the structural element in tension, the scale effect is taken into account using the Weibull approach (Chapter 2, section 2.2).

The tensile strength reference value (i.e. that of the test specimen) corresponds to the loading mode (i.e. a tensile test) under consideration.

If the tensile strength has been determined by means of a splitting or flexure test, it must be corrected accordingly (see section 5.1.5.1 of MC2010 and section 3.1.2 (8) of EC2).

## 3.5. Model accuracy when calculating the strain and crack width

Analytical developments carried out during the CEOS.fr project have shown, especially in the case of thick elements, that taking account of the three-dimensional distribution of strains and stresses, together with a bond-slip law may result in an equation that is more physically exact.

This approach gives acceptable results for the values of crack spacing and crack width.

Additionally, a comparison of the equations and experimental results has shown that the definition of the effective thickness as given in the EC2 and MC2010 codes can give rise to contradictory results for walls.

The current linear two-parameter equation for the crack spacing is considered to be sufficient in simple cases, subject to the above limitations. However, in order to enable better results to be obtained, EC2 should allow for the use of more refined models in order to enable better results to be obtained, especially in the case of thick elements.

PROPOSAL 3.4.– In order to enable a better evaluation of cracking, especially in thick elements, EC2 should allow for the use of more refined models in addition to the current model. These improved models should take account of the three-dimensional distribution of strains, together with a more realistic bond-slip law.

The crack width at the first occurrence of the crack for:

$$\sigma_{s2} = \sigma_{s2,0} = \frac{\gamma+\alpha_e\rho}{\rho} f_{ctm}$$

is given by the equation proposed by Balazs:

$$w_0 = 2g\left(l_{s,max}; \sigma_{s2,0}\right) = 2\left(\frac{(1+\alpha)s_1^{\alpha}\phi f_{ctm}^2}{8\tau_{max}E_s}\left(\frac{\gamma}{\rho}\right)^2\left(1+\frac{\alpha_e\rho}{\gamma}\right)\right)^{\frac{1}{1+\alpha}} \quad [3.7]$$

The crack width once stabilized cracking has been reached is derived from the solution of:

$$w = 2g\left(l_{s,max}, \sigma_{s2}\right) = 2\left(\frac{(1+\alpha)s_1^{\alpha}\phi}{8\tau_{max}E_s\left(1+\frac{\alpha_e\rho}{\gamma}\right)}\sigma_{s2,0}\left(2\sigma_{s2}-\sigma_{s2,0}\right)\right)^{\frac{1}{1+\alpha}} \quad [3.8]$$

This expression simplifies to the previous equation when $\sigma_{s2} = \sigma_{s2,0}$. This equation gives the value of the crack width close to the reinforcement bar. It must therefore be corrected to give the crack width at the surface of the concrete.

The estimated value of the crack width given by [3.8] underestimates the true value, as part of the slippage is not taken into account in the model used in the calculation.

## 3.6. Example of the application of the cracking equations to a concrete tie in tension

This example estimates the cracking (spacing and width) in a concrete tie in tension using the bond-slip law given in MC2010.

The calculation requires a good understanding of the input data, as the end result is very sensitive to these values. In this example, the data has been taken from data obtained from concrete tie tests, supplied to the CEOS.fr project.

The concrete ties consist of beams with a square cross section, 355 × 355 mm, with a length of 3200 mm.

These concrete ties are therefore not considered to be massive. Four reinforcement bars, each 25 mm in diameter, were located in the four corners of the section, with a cover of 65 mm. The ratio of the cross-sectional areas is 1.56 %.

Six concrete ties were tested. The mean strength of the concrete over 18 test specimens was 43.6 MPa, with a low dispersion. The mean tensile strength was 3.99 MPa, with a low dispersion of 4.4%. $\alpha_e$ was taken to be 5.7.

The purpose of this example is to:

– calculate the tensile force applied to the concrete tie to produce the first cracks,

– calculate the mean crack spacing at the stabilised stabilized cracking stage and the mean crack width for an applied tensile force of 500 kN.

The tensile strength of the concrete is affected by the scale effect described in section 2.2.1.

This effect was demonstrated in concrete tests carried out by Pierre Rossi on small test specimens [ROS 94]. The evaluation of this effect was extended to elements with large dimensions following analysis of results from the MEFISTO project (see Chapter 2, section 2.2.1).

NOTE.– In the following example, the scale effect is applied using the method described in Chapter 2, section 2.2.2.

The expression for the mean tensile strength of the concrete tie $f_{ctm}^{Veq}$ is derived from the following equation ([2.3] of section 2.2).

$$\left[\frac{f_{ctm}^{Veq}}{f_{ctm}^{Vref}}\right]^k = \frac{V_{ref}}{V_{eq}}$$

The reference volume $V_{ref}$ is that of the test specimen $V_{specimen}$ used to carry out the tensile strength tests. The test specimen is a cylinder with a diameter of 110 mm and a height of 220 mm, giving a volume $V_{specimen}$ = 2,090,730 mm$^3$. When calculating the Weibull coefficient k, the mean compressive strength has been assumed to be the mean of the values obtained during the tests. Hence:

$$1/k = 0.25 - 3.6 \times 10^{-3} \times 43.6 + 1.3 \times 10^{-5} \times (43.6)^2 = 0.118$$

For first cracking, the volume $V_{eq}$ to be taken into account is the total volume of the concrete tie $V_{tie}$, that is:

$$V_{tie} = 3200 \times 355^2 = 0.403 \text{ m}^3.$$

This volume is less than the maximum volume of 2 m$^3$ specified in section 2.2.2.

In summary:

$$\frac{f_{fctm}^{Veq}}{f_{ctm}^{Vref}} = \left(\frac{V_{tie}}{V_{specimen}}\right)^{-1/k} = \left(\frac{403}{2{,}091}\right)^{-0{,}118} = 0.537 \qquad [3.9]$$

The reference mean tensile strength is that obtained from the test specimens, that is 3.99 MPa. The resulting mean tensile strength of the concrete tie is therefore calculated to be 2.15 MPa.

Given that the concrete ties under consideration do not have a high volume, the coefficient of variation of the standard deviation cannot be considered to be constant, and is to be determined from [2.4]. The value of the exponent B is:

$$B = 4.5 \times 10^{-2} + 4.5 \times 10^{-3} \times 43.6 - 1.8 \times 10^{-5} \times (43.6)^2 = 0.207$$

Hence the dispersion of the tensile strength for the tie:

$$Dispersion_{(tie)} = Dispersion_{(specimen)} * \left(\frac{V_{tie}}{V_{specimen}}\right)^{-B} = 4.4\% \left(\frac{403}{2.091}\right)^{-0.207} = 1.5\%$$

The minimum characteristic value can then be derived as 2.09 MPa.

In order to determine the breaking load for a section during first cracking, the following factors must be taken into rupture account:

– the distribution of stresses in the section prior to cracking (section 3.3.2). This is accounted for using the coefficient $\gamma$ estimated here as 0.7. This gives an apparent tensile strength of 1.46 MPa. This step is the most difficult as, when no values are available, a three-dimensional calculation should be used to calculate $\gamma$ and;

– shrinkage caused by the reinforcement bars, in this case around 0.46 MPa.

Finally, the tensile strength to be taken into account for the first occurrence of cracking is around 1 MPa.

The corresponding tensile force is obtained by multiplying by the homogenized cross-sectional area of the section, giving a value of 138 kN. This value is consistent with observed values. However, these values have a wide variability while remaining well below the value obtained using the value of $f_{ctm}$ given in the standards.

The crack spacing is estimated using the following equation (section 3.3.2):

$$l_{s,max} = \left(\frac{1+\alpha}{1-\alpha}\frac{\gamma f_{ctm}}{4\rho\tau_{max}}\right)^{\frac{1-\alpha}{1+\alpha}} (\phi s_1^{\alpha})^{\frac{1}{1+\alpha}} \left(\frac{2(1-\alpha)^2\left(1+\frac{\alpha_e\rho}{\gamma}\right)}{1+\alpha}\frac{\tau_{max}}{E_s}\right)^{\frac{-\alpha}{1+\alpha}}$$

The values for the parameters in the bond-slip law are those recommended by [DEB 13] based on experimental results obtained by the authors:

$\alpha = 0.25$ $\tau_{max} = 12.5$ MPa; $s_1 = 1.0$ mm

The following comments relate to the tensile strength to be taken into account:

– once the cracking has stabilized (i.e. the last cracks have formed), the volume to be taken into consideration for the scale effect is limited in length to the spacing between two successive cracks whereas, in the case of a concrete tie, the sections are completely cracked. If the crack spacing is taken to be 350 mm, the mean tensile strength becomes 2.78 MPa;

– as a result of the slippage between the reinforcement bars and the concrete, which allows the cracking to stabilize, the restrained shrinkage effect is reduced and may be discounted when calculating the crack spacing; and

– however, the three-dimensional distribution of stresses occurs in a limited volume and the coefficient $\gamma$ is smaller than above.

In summary: $f_{ctm} = 2.78$ MPa; $\gamma = 0.6$.

The calculation gives an anchorage length of 194 mm, corresponding to a maximum crack spacing of 388 mm. In order to obtain the mean spacing, this value must be divided by a coefficient lying between 1.5 and 1.7. Using a value of 1.7 gives: $s_{rm} = 228$ mm. The mean of the 75 spacing measured on the sixties tested was 219 mm compared to the mean spacing's given by MC2010 is 563 mm/1.7 = 331 mm.

The crack width after crack stabilization is given by the following expression (section 3.5):

$$w = 2\left(\frac{(1+\alpha)s_1^{\alpha}\phi}{8\tau_{max}E_s\left(1+\frac{\alpha_e\rho}{\gamma}\right)}\sigma_{s2,0}\left(2\sigma_{s2}-\sigma_{s2,0}\right)\right)^{\frac{1}{1+\alpha}}$$

The initial stress in the reinforcement bar after cracking ($\gamma = 0.7$) is given by:

$$\sigma_{s2,0} = \frac{\gamma + \alpha_e \rho}{\rho} f_{ctm} = 141 \text{ MPa}$$

If the tensile force applied to the concrete tie is 500 kN, that is $\sigma_{s2}$ = 255 MPa, this expression gives a crack width of 392 µm. The mean of the 126 observed values was 131 µm, albeit with a wide variability. The above equation therefore gives a result that is greater than the actual crack width. Slippage is only partially taken into account in this equation [DEB 13]. However, the equation does take into account the maximum value of the crack spacing.

In summary, using an approach to cracking based on the bond-slip laws gives acceptable results for crack spacing, which are better than those predicted by the codes. However, care is required in adopting this approach, as the results are sensitive to the parameters in these laws, and these need to be defined more accurately if the use of this approach is to become the norm.

# 4

# Cracking of Beams Under Mechanical Flexural Loading

Tests carried out on beams by the CEOS.fr project have shown differences between the measured and calculated values of crack spacing. In addition, the crack widths given by EC2 and MC2010 appear to give mean values that are greater than the crack width maximum values (e.g. 95% fractile). Here, we propose that the equations given in these codes should be modified when applied to thick or massive elements.

## 4.1. Crack spacing

$$S_{r,max} = 2l_{s,max} = 2c + \frac{1}{2}\frac{f_{ctm}}{\tau_{bms}}\frac{\emptyset}{\rho_{s,ef}}$$ MC2010 [7.6.4.4]

In this equation, $A_{c,eff.}$ is included as part of $\rho_{s,eff} = A_s/A_{c,eff}$. In the case of beams, the following equation from MC2010 is used to give the value of $A_{c,eff}$:

$$A_{c,eff} = \text{Min}[2.5(h - d)\ ;\ (h - x)/3]$$

NOTE.– There is an error in Figure 7.6.4 of MC2010 in relation to cover c. The cover should be measured from the surface of the reinforcement bar that is closest to the surface of the concrete, and not from the central axis of this bar as shown in the drawing.

## 4.2. Crack width

Two methods of calculating the crack width are proposed:

– calculation of the crack width from the relative strain based on the equations given in MC2010;

– calculation of the crack width by interpolation between the un-cracked stage and the stabilized cracking stage of the beam, applying equations [7.6.14] and [7.6.16] from section 7.6.5.2.2 of MC2010. This method would appear to give improved results in the case of beams.

### 4.2.1. *Tensile stress–strain curve*

The crack width calculations are based on the basic case of a prismatic reinforced concrete tie, subjected to axial tension. The first crack appears at the point R at a tensile stress of $f_{ct,inf}$. As the tensile stress increases, more cracks begin to form up to the point S at which the stabilized cracking stage is reached. This point corresponds to the upper tensile stress $f_{ct,sup}$:

– Figure 4.1(a) is used by both EC2 and MC2010, and assumes that the portion of the curve RS is horizontal. This approximation is considered to be sufficiently accurate in MC2010, resulting in the equality $f_{ct,inf} = f_{ct,sup} = f_{ctm}$;

– Figure 4.1(b) is used when the tensile force increase from $N_{r,fct,inf}$ to $N_{r,fctk,\,0,95}$ is taken into account during the transitional phase between the first crack (point R) and the last crack (point S). This also applies to the stress in the reinforcement bar, which increases from $\sigma_{sr1}$ to $\sigma_{sr2}$. The calculation of the stresses in the reinforcement bars $\sigma_s$ and the corresponding relative strains $\varepsilon_s$ are discussed in detail in section 4.2.3.

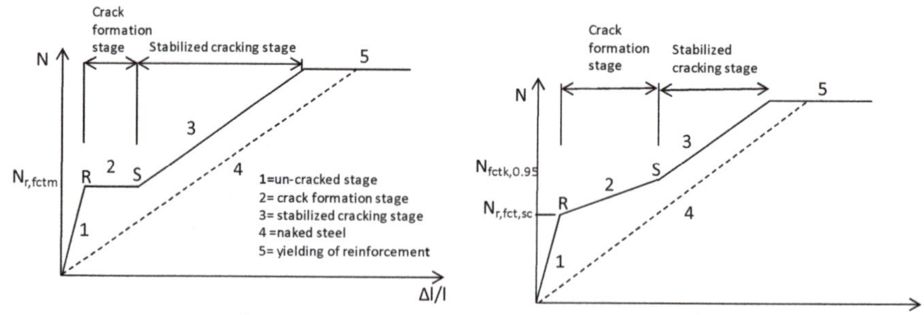


a) b)

**Figure 4.1.** *Stress–strain curves*

### 4.2.2. Calculating the crack width from the relative strain

The crack width is calculated using the following equations taken from MC2010. These equations apply when the stabilized stage is reached under mechanical forces.

$$w_d = 2l_{s,max}(\epsilon_{sm} - \epsilon_{cm})$$ MC2010 section 7.6.4.4.1 Eq. [7.6-3]

$$\varepsilon_{sm} - \varepsilon_{cm} = \frac{\sigma_s - \beta.\sigma_{sr}}{Es}$$ MC2010 section 7.6.4.4.1 Eq. [7.6-5]

where:

$$\sigma_{sr} = \frac{f_{ctm}}{\rho_{s,ef}}(1 + \alpha_e \rho_{s,ef})$$ MC2010 section 7.6.4.4.1 Eq. [7.6.6]

Once stabilized cracking is reached, the crack-free volume of concrete between any two cracks is close to the volume of the reference test specimen used for determining $f_{ctm}$. As a result, the probability of encountering a low strength zone is reduced. Hence, the tensile strength is more likely close to that measured on the specimen. Then it is assumed that the scale effect should be ignored once the stabilized cracking stage is reached at the SLS, unless the crack spacing is significantly greater than the mean dimension of the specimen. In this case, the scale effect should be taken into account as proposed in the method described in section 2.2, applying the scale effect to the value of the mean tensile stress in the concrete $f_{ctm}$.

If the probabilistic scale effect is not applied when estimating the stress $f_{ctm}$ in the concrete after the stabilized cracking stage has been reached, and if the crack spacing is significantly greater than the dimensions of the laboratory test specimen, then the coefficient $\beta$ (see section 3.4) should be multiplied by the factor 0.6. The coefficient $\beta$ is used in the equation for tension stiffening value, $\frac{\beta\sigma_{sr}}{E_s}$. This stiffness reflects the contribution of the concrete between the cracks in accordance with equation [7.6.5] in MC2010 and shown in Figure 4.1(b).

Equation [7.6.5] of MC2010, given above, then becomes:

$$\varepsilon_{sm} - \varepsilon_{cm} = \frac{\sigma_s - 0{,}6\beta.\sigma_{sr}}{Es}$$

As no experimental results are available, the factor 0.6 which multiplies the coefficient $\beta$ is assumed to be still valid as well for short and long term, taking into account the values given in Table 7.6.2 in section 7.6.4.4.1 of MC2010.

### 4.2.3. Calculating the crack width by interpolation between uncracked and fully cracked conditions (the $\zeta$ method)

This method is an alternative way of calculating crack width by interpolation between uncracked and fully cracked conditions.

Equation [7.18] in EN 1992-1-1 [NF 04] is used in the calculation of deflections, and defines the value of a variable in a partially cracked beam by interpolation between the value of the variable in an un-cracked beam (I) and its value in a fully cracked beam (II).

$$\alpha = \zeta\alpha_{II} + (1 - \zeta)\alpha_I \qquad \text{EN 1992-1-1 Eq. [7.18]}$$

where $\alpha_I$ is the effect calculated in the un-cracked state and $\alpha_{II}$ is the effect calculated in the fully cracked state. $\zeta$ is a parameter calculated as a function of the ratio ($\sigma_{sr}/\sigma_s$) given by equation [7.19] of EN 1992-1-1:

$$\zeta = 1 - \beta_\zeta(\frac{\sigma_{sr}}{\sigma_s})^2 \qquad [4.1]$$

where $\beta_\zeta$ is a coefficient taking into account the influence of the duration of the loading or of a repeated loading on the average strain:

– 1.0 for a short-term loading;

– 0.5 for sustained loads or many cycles of repeated loading.

According to MC2010 (section 7.6.5.2.2) and EC2-1, $\sigma_{sr}$ represents the stress in the reinforcement bars in tension, calculated assuming that the section cracked under the loading conditions that are causing the first crack to occur: $\sigma_{sr} = \sigma_{sr1}$. By extension, and to adjust and define the various segments of the curve $\zeta$, $\sigma_{sr}$ is an indexed reference value $\sigma_{sri}$ where i = 1, 2 or 3 (see below).

$\sigma_s$ is the stress in the reinforcement bars in tension, calculated assuming that the cracked section is under the considered loading.

Assuming that the $\alpha$ effect is of a relative strain, $\varepsilon_{sm} - \varepsilon_{cm}$, the above relationship can be written as:

$$\varepsilon_{sm} - \varepsilon_{cm} = \Delta\varepsilon = \zeta\varepsilon_{II} + (1-\zeta)\varepsilon_{II} \qquad [4.2]$$

where $\varepsilon_I$ is the relative strain in the section considered un-cracked, and $\varepsilon_{II}$ is the relative strain in the cracked section.

Unlike the deflections calculation, the effect of creep is taken into account in the calculation of $\varepsilon_I$ and $\varepsilon_{II}$. This is obtained by including the appropriate modulus for the concrete in these calculations, $E_{cm}$ for the short-term calculation and $E_{cm,\ ELSQP}$ for the long-term calculation in a quasi-permanent combination, with:

$$E_{cm,ELSQP} = E_{cm}(1 + \varphi_{ELS,QP})$$

where:

– the creep coefficient $\varphi_{ELS,QP} = \varphi_{\infty}\frac{M_{0Eqp}}{M_{0car}}$;

– $M_{oEqp}$ is the SLS moment in the quasi-permanent combination, and $M_{0car}$ is the SLS moment in the characteristic combination.

The relative strains $\varepsilon_I$ and $\varepsilon_{II}$ are calculated from the following linear equation:

$$\varepsilon = (z - x)\, M/EI \qquad [4.3]$$

where z – x is the distance from the neutral axis to the fiber for which the value of $\varepsilon$ is required, where:

– x is the depth of the neutral axis;

– z is the depth of the fiber for which $\varepsilon$ is calculated;

– M is the flexural moment.

The value of $\varepsilon_I$ is calculated using the values of x, E, I and M corresponding to the uncracked stage. The value of $\varepsilon_{II}$ is then calculated using these values in the cracking stage.

The value of $\varepsilon_I$ and the value of $\varepsilon_{II}$ are obtained using the respective values of the moment $M_{cr1}$ and $M_{cr2}$, in [4.3].

The scale effect is introduced by the stress $\sigma_{sr}$, used in the equation $\zeta = 1 - \beta_\zeta (\frac{\sigma_{sr}}{\sigma_s})^2$:

$$\sigma_{sr} = \frac{f_{ctm}}{\rho_{s,ef}}(1 + \alpha_e \rho_{s,ef})$$

from MC2010 and depending on $f_{ctm}$.

It should be noted that the value of $f_{ctm}$ varies according to the crack formation stage (RS line 2 in Figure 4.1(b)). As a result, the value of $f_{ct}$ to be used in the above equation varies according to the extent of the cracking.

The value of $f_{ct}$ corresponding to the formation of the first crack gives the point of intersection $R(\sigma_{sr1}, \varepsilon_{sr1})$ of the lines 1 and 2 in Figure 4.1. This value is assumed to be equal to $f_{tsc}$, the lowest value of $f_{ct}$ in the structure calculated by applying an approach based on Weibull's theory (see section 2.2).

$$f_{t,sc} = f_{ct,0.05}^{V_{ref}} = f_{ctk,0.05}^{V_{ref}}(V_{ref}/V_{eq})^{1/k_V}$$

where $k_v$ is the Weibull exponent and $f_{cm}$ is the mean compressive strength (Table 3.1 in EN 1992-1-1):

$$1/k_v = 0.25 - 3.6\ 10^{-3}\ f_{cm} + 1.3\ 10^{-5}\ f_{cm}^2$$

and the volumes are:

$$V_{eq} = \text{Min}(1.25;l).\text{Min}(1.25;b).\text{Min}(1.25;h_{cef})$$

where:

– $V_{ref}$ is the volume in tension in the tensile specimen, for the split test $V_{ref} = 0.0003\ m^3$.

– $V_{eq}$ is the volume of the equivalent tie in the beam (length l, width b and height $h_{c,ef}$ as defined in EC2-1-1 and MC 2010: $h_{c,ef} = \text{Min}(2.5(h–d);(h–x)/3)$, but if the minimum is (h–x)/3 and the reinforcement is not within (h–x)/3; 2.5(h–d) should be adopted.

The value of $f_{tsc}$ is used to calculate the moment $M_{cr1}$ in the homogeneous section by the equation:

$$M_{cr2} = f_{ctk,0.95}\, I_{hom}/(h - x_{hom})$$

From this bending moment corresponding to the first crack formation, it is possible to derive from $M_{crl}$ the stress in the reinforcement bar in the section after cracking:

$$\sigma_{srl} = \alpha_e M_{crl} (d - x_{fis})/ I_{fis},$$

where:

– x is the depth of the neutral axis;

– I is the inertia of the homogeneous section (index $_{hom}$) and the cracked section (index $_{fis}$);

– $\alpha_e = E_s / E_{cm,SLS}$

In this section, the strain in the homogenous structure $\varepsilon_{srl}$ is known.

At the point S ($\sigma_{sr2}$, $\varepsilon_{sr2}$), corresponding to the last crack formation, the stress $f_{ct,w}$ is the stress in the reinforcement bar in the section after cracking, equal to the maximum value of $f_{ct}$, that is $f_{ctk,\ 095}$. This point cannot be calculated directly with the above method as the relationship $\zeta = 1 - \beta_\zeta (\sigma_{sr} / \sigma_s)^2$ results in the value of $\zeta$ being zero. The strain $\varepsilon_{sr2} = \varepsilon_{sm} - \varepsilon_{cm}$ will be calculated using the equation from MC2010 as shown below.

The value $f_{ctk,\ 0,95}$ is used to calculate the corresponding moment $M_{cr2}$:

$$M_{cr2} = f_{ctk,\ 0,95}\ I_{hom} / (h - X_{hom})$$

This corresponds to the last crack formation. Similarly, this can be used to calculate:

$$\sigma_{sr2} = \alpha_e M_{cr2}\,(d - x_{fis})/I_{fis} \qquad [4.4]$$

which is the stress in the reinforcement bar in the section after cracking,

When calculating $\varepsilon_{sr2}$ at the point S, a value derived from experimentation is close to that obtained from equation [7.6.5] in section 7.6.4.4.1 of MC2010 with the value of β specified in this code.

$$\varepsilon_{sr2} = \frac{\sigma_{sr2} - \beta.\sigma_{sr}}{Es} \qquad \text{MC2010 section 7.6.4.4.1 Eq. [7.6-5]}$$

and

$$\sigma_{sr} = \frac{f_{ctm}}{\rho_{s,ef}}(1 + \alpha_e \rho_{s,ef})$$

Then, $\zeta_{sr2}$ can be calculated by the relationship $\varepsilon = \zeta\varepsilon_{II} + (1-\zeta)\varepsilon_I$ for the value of $\varepsilon_{sr2}$:

$$\zeta_{sr2} = (\varepsilon_{sr2} - \varepsilon_I)/(\varepsilon_{II} - \varepsilon_I)$$

where $\varepsilon_{II}$ and $\varepsilon_I$ are the strains of the fully cracked and of the uncracked sections, respectively, as calculated using [4.3].

The values of $\zeta$ can then be interpolated between $\sigma_{sr1}$ and $\sigma_{sr2}$:

$$\zeta_\sigma = \zeta_{sr2}\left[1 - (\sigma_{sr2} - \sigma)/(\sigma_{sr2} - \sigma_{sr1})\right]$$

$\zeta_\sigma = 0$ for $\sigma \leq \sigma_{sr1}$.

For $\sigma > \sigma_{sr2}$, the $\zeta$ curve passing through $\zeta_{sr2}$ should be determined. The corresponding $\sigma_{sr3}$ is:

$$\sigma_{sr3} = \sigma_{sr2}\sqrt{\frac{1-\zeta_{sr2}}{\beta_\zeta}}$$

then:

$$\zeta_\sigma = 1 - \beta_\zeta(\sigma_{sr3}/\sigma_s)^2$$

for all values of $\sigma$ (except $\sigma_{sr2}$ calculated using [4.4]), using a value of $\beta_\zeta = 1$ for short-term loading or 0.5 for long term or cyclical loading over a large number of cycles.

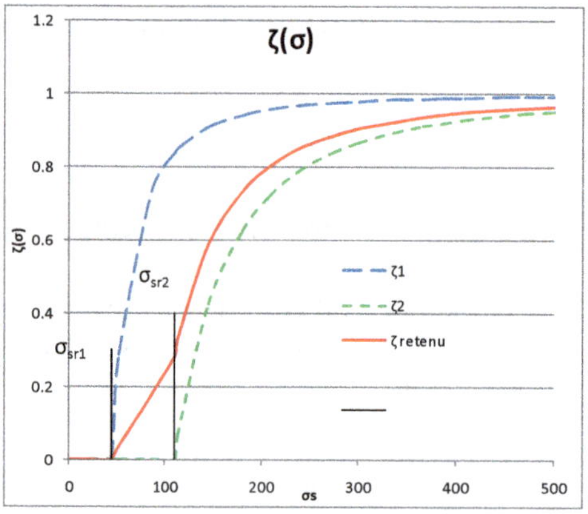


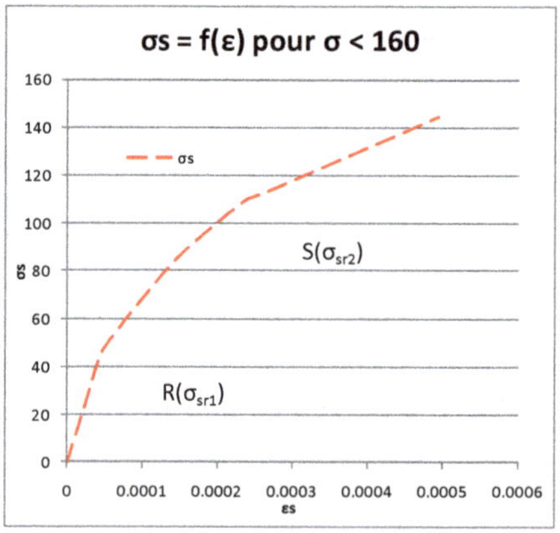


**Figure 4.2.** *Extract from CEOS.fr: Stress/strain relationship for beam (left) for comparison with Figure 4.1(b). The function $\zeta(\sigma)$ (right), where $\zeta_1$ and $\zeta_2$ are calculated from $\sigma_{sr1}$ and $\sigma_{sr2}$ respectively*

SUMMARY.–

The values of $\zeta$ according to $\sigma$ are given in Table 4.1 and Figure 4.3 below.

| State | $\sigma_s$ | $\zeta_\sigma$ |
|---|---|---|
| Uncracked | $< \sigma_{sr1}$ | 0 |
| First crack | $\sigma_{sr1} = \alpha_e M_{cr1} (d - x_{fis})/I_{fis}$ | $1 - \beta_\zeta (\sigma_{sr1} / \sigma_s)^2 = 0$ |
| Progressive cracking | $\sigma_{sr1} < \sigma < \sigma_{sr2}$ | $\zeta_\sigma = \zeta_{sr2} [1 - (\sigma_{sr2} - \sigma_s) / (\sigma_{sr2} - \sigma_{sr1})]$ |
| Last crack | $\sigma_{sr2} = \alpha_e M_{cr2} (d - x_{fis})/I_{fis}$ | $\zeta_{sr2} = (\varepsilon_{sr2} - \varepsilon_I) / (\varepsilon_{II} - \varepsilon_I)$, |
| Stabilized crack stage | $> \sigma_{sr2}$ | $\zeta_\sigma = 1 - \beta_\zeta (\sigma_{sr3} / \sigma_s)^2$ where $\sigma_{sr3} = \sigma_{sr2}\sqrt{\frac{1-\zeta_{sr2}}{\beta_\zeta}}$ |

**Table 4.1.** *Values of $\zeta$ according to $\sigma$*

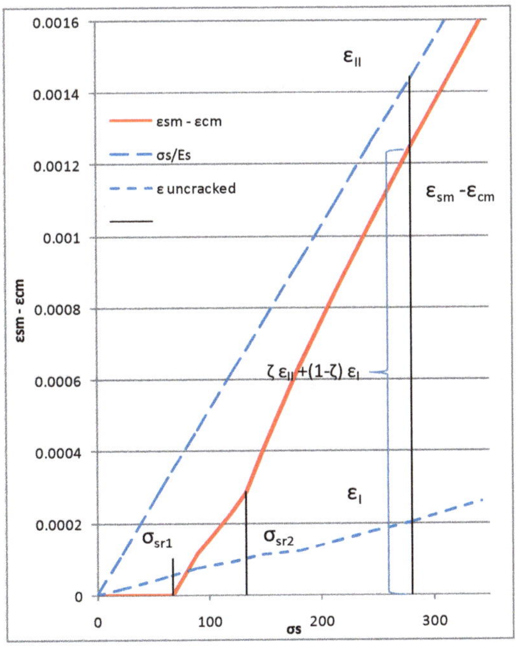


**Figure 4.3.** *Diagram of $\varepsilon_{sm} - \varepsilon_{cm}$ using interpolation method*

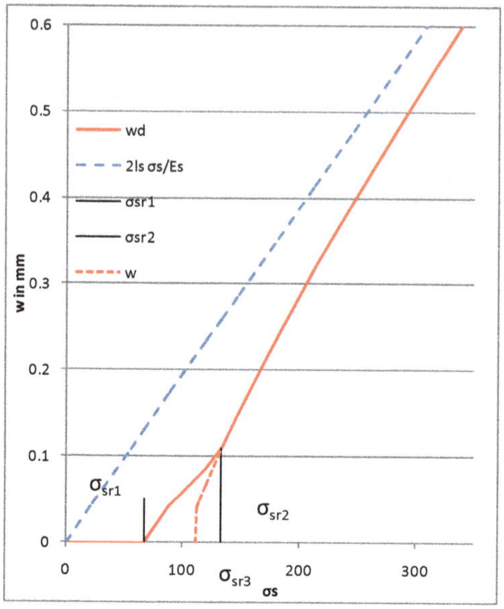


**Figure 4.4.** *Diagram of $w_d$ using interpolation method*

The value of $\varepsilon_{sm} - \varepsilon_{cm}$ for each value of $\sigma_{sr}$ is given by:

$$\varepsilon(\sigma_s) = (\varepsilon_{sm} - \varepsilon_{cm})_\sigma = \zeta_\sigma \varepsilon_{II}(\sigma_s) + (1 - \zeta_\sigma)\varepsilon_I(\sigma_s)$$

The advantage of this approach is to take into account the slope of line 2 in Figure 4.1(b), unlike the method given in MC2010 in which this line is assumed to be horizontal in order to simplify the calculation (Figure 4.1(a)).

For values of $\sigma_s > \sigma_{sr1}$ (i.e. when $\sigma_s \leq \sigma_{sr1}$, $w_d = 0$), the crack width $w_d$ is calculated using equation [7.6.3] of MC2010), using the values of $(\varepsilon_{sm} - \varepsilon_{cm})$ calculated above:

$$w_d = 2l_{s,max}(\epsilon_{sm} - \epsilon_{cm}) \qquad \text{MC2010 Eq. [7.6-3]}$$

where:

$$l_{s,max} = c + \frac{1}{4}\frac{f_{ctm}}{\tau_{bms}}\frac{\emptyset}{\rho_{s,ef}}$$

Figure 4.3 shows the evolution of $\varepsilon_{sm} - \varepsilon_{cm}$ with $\sigma_s$ and Figure 4.4 the corresponding evolution of $w_d$.

## 4.3. Examples

### 4.3.1. *Example 1: calculation of crack spacing and crack width in a thick concrete slab under heavy loads*

Formwork:

– Slab with a length of l = 12.00 m, thickness h = 1.20 m, and width b = 3.00 m.

Materials:

– Concrete: $f_{ck}$ = 40 MPa, $f_{ctm}$ = 3.5 MPa (measured by direct tension), $E_{cm}$ = 35,000 MPa.

– Steel: $f_{yk}$ = 500 MPa, $E_s$ = 200,000 MPa, $\Phi$ = 32 mm ($\Phi_{ext}$ = 35 mm).

Environment:

– Exposure class: XS1, i.e. $c_{nom}$ = Max ($\Phi$+10; 30+10) in mm.

– Cracking: $w_d$ = 0.2 mm.

Loading and corresponding efforts:

– Own weight: G = 90 kN/ml;

– Working load: 80 kN/m² i.e. Q = 240 kN/ml with $\psi_2$ = 0.7 (coefficient for a quasi-permanent combination);

This gives the following forces:

– ELU $M_{ELU}$ = 8.667 MNm and $T_{ELU}$ = 2.889 MN;

– Characteristic SLS $M_{ELScar}$ = 5.94 MNm and $T_{ELScar}$ = 1.98 MN;

– Quasi-permanent SLS $M_{ELSqp}$ = 4.644 MNm and $T_{ELS}$ = 1.548 MN.

Reinforcement:

– Minimum percentage of 9.2.1.1 $A_{smin}$ = 61.92 cm²;

– ULS reinforcement $A_{sELU}$ = 176.9 cm² (22 $\Phi$32);

– Quasi-permanent SLS reinforcement for $w_d$ = 0.2 mm $A_{s,ELSqp}$ = 233.16 cm²;

– i.e. First layer 22 $\phi$32 and second layer 7 $\phi$32 and total 29 $\phi$32;

– d = 1.134 m.

Characteristics of sections:

– Cracked section at quasi-permanent SLS;

– $\varphi_\infty = 2$, $\varphi_{ef,ELS} = 1.6$, $E_{cm,ELS} = 13{,}652$ MPa;

– hence $\alpha_e = n = E_s / E_{cm,ELS} = 14.6$;

– This gives $x_{fis} = 0.384$ m, $I_{fis} = 0.2818$ m$^4$, and $\sigma_s = 180.5$ MPa;

– Homogeneous section at quasi-permanent SLS;

– $x_{hom} = 0.63$ m and $I_{hom} = 0.5556$ m$^4$.

Cracking calculation under long term loading:

*According to MC2010*;

– $h_{cef} = 0.165$ m, $A_{cef} = 0.495$, $\rho = 0.047103$;

– With $\beta = 0.4$, this gives $\sigma_{sr} = 94.3$ MPa and $2\ l_{s,max} = 273$ mm;

– i.e. $\varepsilon_{sm} - \varepsilon_{cm} = 0.0007138$, which gives $w_{MC2010} = 0.195$ mm;

*MC2010; 0.6 in accordance with Section 4.2.2*;

– $\varepsilon_{sm} - \varepsilon_{cm} = 0.0007892$ which gives $w_{MC2010;0.6} = 0.215$ mm.

Using the $\zeta$ method in accordance with section 4.2.3:

– Calculation of $f_{tsc}$ (lowest value of $f_t$ found across the height $h_{cef}$ of the beam):

- $f_{ctm,sp} = f_{ctm} / 0.9$ (tensile strength measured by a splitting test) = 3.89 MPa;

- Vref = 0.0003, Veq = 0.248, 1/m = 0.10715, hence $f_{tsc} = 1.89$ MPa;

– Calculation of $\zeta$ (remember $\beta_\zeta = 0.5$ long term):

- $M_{cr1} = f_{tsc}\ I_{hom} / (h - x_{hom}) = 1.85$, hence $\sigma_{sr1} = n\ M_{cr1}\ (d - x_{fis}) / I_{fis} =$ 71.7 MPa;

- $M_{cr2} = 1.3\ f_{ctm}\ I_{hom} / (h - x_{hom}) = 4{,}45$, hence $\sigma_{sr2} = n\ M_{cr2}\ (d - x_{fis}) / I_{fis} =$ 172.7 MPa;

- $\varepsilon_{sr2}$ (from MC2010) = $(\sigma_{sr2} - \beta\ \sigma_{sr})/E_s$ ($\beta = 0.4$ and $\sigma_{sr} = 94.25$);

– Strain under $M_{cr2}$:

- Homogeneous section: $\varepsilon_1 = M_{cr2}\,(d - x_{hom}) / (E_{cmELS}\,I_{hom}) = 0.0002946$;
- Cracked section: $\varepsilon_2 = M_{cr2}\,(d - x_{fis}) / (E_{cmELS}\,I_{fis}) = 0.0008666$;
- Hence $\zeta_{sr2} = (\varepsilon_{sr2} - \varepsilon_1) / (\varepsilon_2 - \varepsilon_1) = 0.6651$;
- and $\sigma_{sr3} = \sigma_{sr2}\,(1 - \zeta_{sr2})$^0,5 = 141.35 MPa.

– Calculating the cracking in the beam under $\sigma_s = 180.5$ MPa:

- as $\sigma_s > \sigma_{sr2}$, $\zeta_\sigma = \zeta_{sr2}\,(1 - \beta_\zeta\,(\sigma_{sr3} / \sigma_s)^2) = 0.6932$;

– Calculating the strains under $\sigma_s$:

- $\varepsilon_1(\sigma_s) = M_{ELSqp}\,(d - x_{hom}) / (E_{cmELS}\,I_{hom}) = 0.00030778$;
- $\varepsilon_2(\sigma_s) = M_{ELSqp}\,(d - x_{fis}) / (E_{cmELS}\,I_{fis}) = 0.00090534$.

– Resulting strain from the equation $\varepsilon_{sm} - \varepsilon_{cm} = \Delta_\varepsilon = \zeta\,\varepsilon_{II} + (1 - \zeta)\,\varepsilon_I$:

- $\varepsilon_{sm} - \varepsilon_{cm} = 0.0007220$.

Hence:

$\mathbf{w_\zeta = 2l_{smax}\,(\varepsilon_{sm} - \varepsilon_{cm}) = 0.197\ mm}$

### 4.3.2. *Example 2: calculation of crack spacing and crack width in a double thick beam*

Formwork:

– Double T beam, span l = 12.00 m;

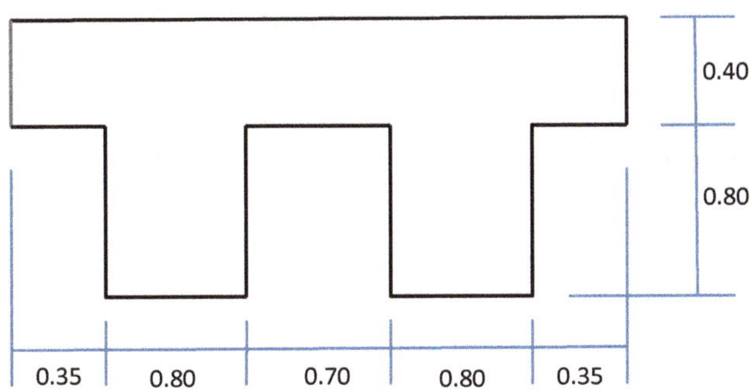

Materials:

– Concrete: $f_{ck}$ = 40 MPa, $f_{ctm}$ = 3.5 MPa (measured by direct tension), $E_{cm}$ = 35 000 MPa;

– Steel: $f_{yk}$ = 500 MPa, $E_s$ = 200 000 MPa, $\Phi$ = 32 mm ($\Phi_{ext}$ = 35 mm).

Environment:

– Exposure class: XS1, i.e. $c_{nom}$ = Max ($\phi$+10; 30+10) in mm = 42 mm.;

– Cracking: $w_d$ = 0.2 mm.

Loading and corresponding efforts:

– Own weight: G = 62 kN/ml;

– Working load: 100 kN/m² i.e. Q = 300 kN/ml with $\psi_2$ = 1 (coefficient for a quasi-permanent combination).

This gives the following forces:

– ULS $M_{ELU}$ = 9.61 MNm and $T_{ELU}$ = 3.20 MN;

– Characteristic SLS $M_{ELScar}$ = 6.52 MNm and $T_{ELScar}$ = 2.17 MN;

– Quasi-permanent SLS $M_{ELSqp}$ = 6.52 MNm and $T_{ELSqp}$ = 2.17 MN.

Reinforcement:

– Minimum percentage of 9.2.1.1 $A_{smin}$ = 32.8 cm$^2$;

– ULS reinforcement $A_{sELU}$ = 201 cm$^2$ (25 $\Phi$32);

– Quasi-permanent SLS reinforcement for $w_d$ = 0.2 mm $A_{s,ELSqp}$ = 265.32 cm²;

– i.e. first layer 18 $\Phi$32, second layer 15 $\Phi$32, and total 33 $\Phi$32;

– d = 1.126 m.

Characteristics of sections:

– $\varphi\infty$ = 2, applying Section section 4.2.3 above $\varphi_{ef,ELS}$ = 2.0, $E_{cm,ELS}$ = 11,667 MPa;

Hence $\alpha_e$ = n = Es / $E_{cm,ELS}$ = 17.1:

– This gives $x_{fis}$ = 0.431m, $I_{fis}$ = 0.33603 m$^4$, and $\sigma_s$ = 230 MPa;

– Homogeneous section at quasi-permanent SLS;

– $x_{hom} = 0.59$ and $I_{hom} = 0.481740\ m^4$.

Cracking calculation under long term loading:

*According to MC2010*:

– $h_{cef} = 0.185$ m, $A_{cef} = 0.296$ , $\rho = 0.089635$;

– With $\beta = 0.4$, this gives $\sigma_{sr} = 59.00$ MPa and 2 $I_{s,max} = 183$ mm;

– i.e. $\varepsilon_{sm} - \varepsilon_{cm} = 0.00103426$;

– $w_{MC2010} = 0.189$ *mm*.

*MC2010; 0.6 in accordance with section 4.2.2*:

– $\varepsilon_{sm} - \varepsilon_{cm} = 0.00108146$;

– $w_{MC2010;0,6} = 0.198$ *mm*;

– *Using the* $\zeta$ *method in accordance with section 4.2.3.*

Calculation of $f_{tsc}$ (lowest value of ft found across the height $h_{cef}$ of the beam):

– $f_{ctm,sp} = f_{ctm} /0.9$ (tensile strength measured by a splitting test) = 3.89 MPa;

– $V_{ref} = 0.0003$, $V_{eq} = 0.289$, $1/m = 0.107152$, hence $f_{tsc} = 1.86$ MPa.

Calculation of $\zeta$ (remember $\beta_{\zeta} = 0.5$ long term):

– $M_{cr1} = f_{tsc}\ I_{hom} / (h-x_{hom}) = 1.464$, hence $\sigma_{sr1} = n\ M_{cr1}\ (d - x_{fis}) / I_{fis}$ = 51.8 MPa;

– $M_{cr2} = 1.3\ f_{ctm}\ I_{hom} / (h-x_{hom}) = 3.58$, hence $\sigma_{sr2} = n\ M_{cr2}\ (d - x_{fis}) / I_{fis}$ = 126.5 MPa;

– $\varepsilon_{sr2}$ (in accordance with MC2010) = $(\sigma_{sr2} - \beta\ \sigma_{sr})/E_s$( $\beta = 0,4$ and $\sigma_{sr} = 59$ MPa);

– $\varepsilon_{sr2} = 0.000514506$.

Strain under $M_{cr2}$:

– Homogeneous section: $\varepsilon_1 = M_{cr2}\ (d - x_{hom}) / (E_{cmELS}\ I_{hom}) = 0.0003429$;

– Cracked section: $\varepsilon_2 = M_{cr2}$ (d – $x_{fis}$) / ($E_{cmELS}$ $I_{fis}$) = 0.0006342;

– Hence $\zeta_{sr2} = (\varepsilon_{sr2} - \varepsilon_1)/(\varepsilon_2 - \varepsilon_1) = 0.5891$;

– and $\sigma_{sr3} = \sigma_{sr2}$ (1 - $\zeta_{sr2}$)^0,5 = 128.21 MPa.

Calculating the cracking in the beam under $\sigma_s = 230$ MPa:

– as $\sigma_s > \sigma_{sr2}$, $\zeta_\sigma = \zeta_{sr2} (1 - \beta_\zeta (\sigma_{sr3} / \sigma_s)^2) = 0.87619$.

Calculating the strains under $\sigma_s$:

– $\varepsilon_1(\sigma_s) = M_{ELSqp}$ (d – $x_{hom}$) / ($E_{cmELS}$ $I_{hom}$) = 0.000624561;

– $\varepsilon_2(\sigma_s) = M_{ELSqp}$ (d – $x_{fis}$) / ($E_{cmELS}$ $I_{fis}$) = 0.001155145.

Resulting strain from the equation $\varepsilon_{sm} - \varepsilon_{cm} = \Delta\varepsilon = \zeta\varepsilon_{II} + (1-\zeta)\varepsilon_{II}$:

– $\varepsilon_{sm} - \varepsilon_{cm} = 0.0010895$.

Hence:

– **$w_\zeta = 2l_{smax} (\varepsilon_{sm} - \varepsilon_{cm}) = 0.199$ mm**

In this second example, the ordering of the respective crack width values for MC2010;0,6 and $\zeta$ method is reversed as the values of $\sigma_s$ are greater than the corresponding value at the point of intersection of the MC2010;0,6 curve and the $\zeta$ curve.

# 5

# Cracking in Walls

The following recommendations relate to the calculation of cracking in walls subject to a shear force (bracing), and may also be applied to beam webs and caissons. First, walls subjected to quasi-static forces are considered. In the second part, cracking under the effect of earthquakes is also addressed.

## 5.1. Current status of the reference texts

EN 1992-1-1 section 7.3.4 text is given below:

> (4) Where the angle between the axes of principal stress and the direction of the reinforcement, for members reinforced in two orthogonal directions, is significant (>15°), then the crack spacings $R_{,max}$ may be calculated from the following expression:
>
> $$S_{r,max} = \frac{1}{\frac{\cos\theta}{S_{r,max,y}} + \frac{\sin\theta}{S_{r,max,z}}}$$
>
> where:
>
> – $\theta$ is the angle between the reinforcement in the y direction and the direction of the principal tensile stress, and
>
> – $s_{r,max,y}$ and $s_{r,max,z}$ are the crack spacing calculated in the y and z directions respectively, according to 7.3.4 (3).

Model Code 2010 adopts a similar equation, but with the slippage lengths $l_{s,x}$ and $l_{s,y}$ in place of the crack spacing, and with the length $l_{s,max,\theta}$ (in place of $s_{r,max}$) perpendicular to the crack being virtual as there is no reinforcement in that direction.

The application of the equation given in EN 1992-1-1 and MC2010 raises a number of unanswered questions, which require a response. In particular, there is no mention of how to derive crack width from the expression for crack spacing. From this perspective, MC2010 begins to offer a solution by adding the following to the EC2 clause as follows:

The design crack width can then be calculated from:

$$w_d = 2 \cdot l_{s,max,\theta}(\varepsilon_\perp - \varepsilon_{c,\perp})$$

where:

– $\varepsilon_\perp$ and $\varepsilon_{c,\perp}$ represent the total mean strain and the mean concrete strain, evaluated in the direction orthogonal to the crack (see Figure 5.1).

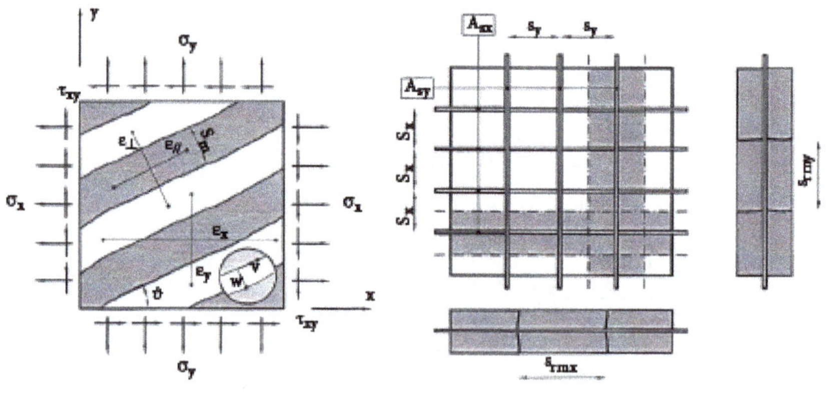


**Figure 5.1.** *Calculation model suggested by MC2010*

In a concrete tie, the mean total strain is equal to the mean strain of the reinforcement bars. This relationship may be applied, however it cannot be justified, as explained in section 5.2.

The proposed method is therefore based on a physical "strut and tie" model, as detailed by several authors in sophisticated models such as the Cracked Membrane Model (CMM) [KAU 98]. No practical instructions for applying this method are given in MC2010.

## 5.2. Validity of the physical model and calculating of the crack angle

The physical model to which EN 1992-1-1 and MC2010 refer to is a "strut and tie" type and is not valid for all stress situations. Such a model is only applicable if the principal stresses are one tensile stress and one compressive stress. In particular, this excludes the case where both stresses are tensile. In addition, it is understood that cracking in a wall may be horizontal if only the concrete is resisting the shear force, in which case it is the bending moment that determines the cracking, hence the strut and tie model is not valid.

Inclined cracking therefore implies that the shear stress in the horizontal plane exceeds a certain criterion, for example, the principal stress exceeds the tensile strength. However, the criterion may be more general as in the case, for example, of the shear resistance model in MC2010 and EC2, where shear reinforcement is only necessary if shear force is greater than $V_{RdC}$.

When a load combination results in cracking at a given point, both MC2010 and EC2 indicate that the crack will be orthogonal to the direction of the principal tensile stress. This approximation is adequate in practice, as under certain conditions (see section 5.4) crack evaluation has limited sensitivity to variations in the angle. In practice, the membrane forces in the plane of a wall are determined by a structural analysis. For simplification, it is assumed that the membrane forces are given in the reference plane defined by the $x$ (horizontal) and $y$ (vertical) axes, which are also the same directions as the reinforcement bars in the wall. This method can be further generalized to include other axis systems, but this is not presented here for an easier understanding.

For any combination of loads, the resultant shear force applied to a plane with an angle $\theta$ to the horizontal $x$ (i.e. tangential to the crack) is given by the following equation (see section 5.5):

$$T_r = (N_{yy} - N_{xx}) + sin\theta cos\theta(cos^2\theta - sin^2\theta)$$

$T_r$ is zero for the combination that creates the crack, assuming that the crack is traction-free, and this determines $\theta$ (see section 5.5). Once the crack has formed, its angle does not change locally, and $\theta$ is fixed for other combinations of actions. This is especially the case when the shear stress increases, as in an earthquake. As $\theta$ is fixed, $T_r$ is not zero when $N_{xy}$ increases (Note: there are some calculation models in which the crack turns, including the CMM cited above). However, the state of the stresses is different at each point in the wall, and therefore the cracking does not occur simultaneously across the entire wall and the angle may vary from one point to another.

Inclined cracking, associated with a strut and tie physical model, is only valid if the shear force is not resisted only by the concrete and if the principal stresses are one tensile stress and one compressive stress. The crack angle is then determined, to an adequate approximation, by the direction orthogonal to the principal direction of the tensile stress for the combination of loads that creates the crack.

## 5.3. Calculation model

The state of strain and stresses in a wall rarely result in a "simple" calculation, and structural effects usually dominate in determining strain and stress distribution. It is therefore necessary to use either a traditional wind bracing calculation or finite element modeling in order to evaluate the stresses and strains. The results may be used to:

– assess the validity of the strut and tie model (see section 5.2);

– calculate the angle of inclination of the cracks (see section 5.2);

– calculate the mean relative strains (see section 5.5).

The quality of the modeling is therefore a crucial factor in providing a realistic evaluation of the cracking. The intention here is not to discuss the full range of good practice involved in creating a representative finite element model to the required degree of accuracy. Rather it is simply to draw attention to the significance of the wall behavior constitutive model:

– most models give results which enable the crack angle to be evaluated to a sufficient degree of accuracy, providing that the reinforcement is close to optimum (see section 5.4);

– the non-linear models for the behavior of reinforced concrete typically give highly variable results when used to evaluate strains. If the results, in terms of the mean strain, are sufficiently reliable locally, then these can be used directly in calculating the crack width. However, this type of model may give good results when evaluating the overall behavior of the structure without necessarily producing an accurate local representation of strains;

– the traditional linear elastic calculation typically gives acceptable results for the stress distribution, but not for strains.

Section 5.5 provides a range of methods for calculating the mean relative strains as a function of the stresses. It is therefore a minimum requirement that any calculation model should adequately represent the stress distribution.

## 5.4. Crack spacing and slippage length

The calculation of crack spacing in walls subject to membrane shear forces, which create through wall cracking, is based on the use of virtual ties in the vertical and horizontal directions.

This approach is given in EC2 and MC2010, and in more sophisticated models for cracking in walls. Hence, the relevance of any model for cracking in walls is based partly on the applicability of the model for cracking in concrete ties, and partly on the validity of the applied equation developed by [VEC 86]:

$$S_{rm} = \left(\frac{\sin\theta}{S_{rmx}} + \frac{\cos\theta}{S_{rmy}}\right)^{-1}$$

The components $s_{rmx}$ (horizontal direction) and $s_{rmy}$ (vertical direction) are crack spacing in the two virtual ties, taking account of the reinforcement bars in each direction. This equation may be used for both "maximum" values, as in EC2, or mean values. In MC2010, the slippage lengths are considered in place of the crack spacing. This enables the method to include the cracking formation phase.

$s_{rmx}$ and $s_{rmy}$ must be evaluated accurately using equations suitable for concrete ties (see section 3.3). In addition, the way in which the various values of the cover $c$ for the two reinforcement layers on either side of the wall are taken into account is not discussed in either EC2-1-1 or MC2010.

*When calculating the crack spacing in the horizontal x and vertical y directions in walls containing two layers of reinforcement bars using EC2 equations (or the slippage length from MC2010), it is proposed that the mean cover of the two layers should be used. This approach is recommended, alternatively each layer should use its own value for cover.*

The Vecchio and Collins equation initially assumes that the strut and tie model is physically valid (see section 5.2). $s_{rmx}$ and $s_{rmy}$ are the geometric quantities defined in Figure 5.2. The assumption is based on the fact that these are the crack spacing in the virtual ties in the $x$ and $y$ directions of the reinforcement bars. This assumption has not been confirmed by the results of tests carried out on walls as part of the CEOS.fr project. Several authors have also questioned it. However, it does result in reasonable values (when compared to experimental results) under the conditions described below.

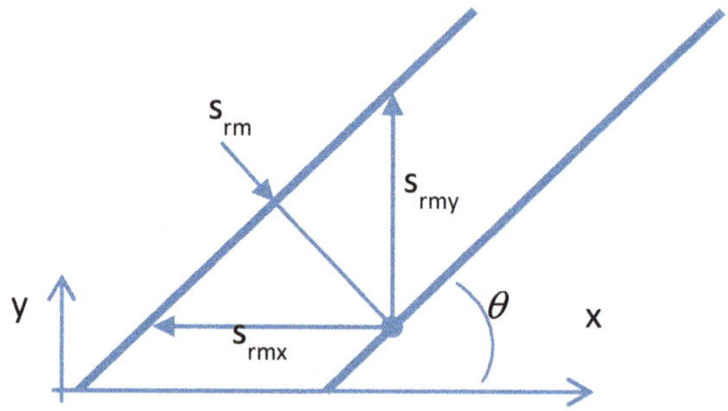


**Figure 5.2.** *Crack spacing in the virtual tie in x and y directions of reinforcement bars*

This equation may be used as a first approximation, although it does however require further analysis. An initied comment relates to consideration of the optimum reinforcement from the point of view of cracking. For the combination of loads creating the crack at a given point, and $\theta$ being determined as described in section 5.2, the crack width is a minimum when the reinforcement bars in the $x$ and $y$ directions are in the proportions such that:

$$\frac{s_{rmx}}{s_{rmy}} = cotg\theta \qquad [5.1]$$

If this optimum condition has been met, the value of the crack spacing shows little sensitivity to variations in the angle $\theta$ resulting from the structural analysis. A further result is that the Vecchio and Collins equation [KAU 98] is acceptable within a range of crack angles of approximately ± 20° around an angle defined by:

$$\theta = arccotg\left(\frac{s_{rmx}}{s_{rmy}}\right) \qquad [5.2]$$

There is a clear consistency between the two results. Whilst this approach is optimum from the point of view of cracking, it is not necessarily optimum when considering the reinforcement bar dimensions for strength, which usually takes into account a different combination of loads at the ULS.

A second comment, although the equation is clearly no longer applicable under these conditions for reasons of geometry, is that when the value of $\theta$ is low (or close to 90°) one term in the Vecchio and Collins equation is small when compared to the other, due to sin $\theta$ (or cos $\theta$), and has no effect on the result. Finally, the (mean) values obtained from the Vecchio and Collins equation are comparable with experimental results obtained by CEOS.fr where the crack angle is around 30°, compared with an optimum value of around 45°.

*The Vecchio and Collins equation may be used as a reasonable approximation within ± 20°of the optimum angle determined for the reinforcement bars, provided that the strut and tie model remains valid.*

A number of other results from the CEOS.fr research project are considered appropriate for practical applications, these include that:

– the first cracks may appear at level of tensile stress well below the tensile strength, even when the behavior of early age concrete is taken into account. There are a number of phenomena that may explain this difference, including the variability of the concrete strength and the scale effect, the distribution of stresses throughout the thickness of the wall, and shrinkage restrained by the reinforcement bars, etc.;

– contrary to the EC2 equation, the crack spacing is dependent on the strength of the concrete. The higher the strength, the wider the crack spacing. There is no obvious explanation for this observation based on the state of knowledge at this point in time;

– when compared to the experimental results, the definition of the effective thickness, as given in EC2 and MC2010 gives rise to inconsistencies in the case of walls. Considering the total thickness instead of the effective thickness may improve the prediction for thick walls, although the resulting answer is not completely satisfactory;

– MC2010 calculates the crack spacing values from the slippage lengths, which are more realistic than the usually overestimated spacing values given by EC2 [7.11]. This difference was noted in the case of concrete ties (section 3.3.1) and is also the case for walls.

## 5.5. Mean differential strain

EC2-1-1 does not consider methods for calculating mean differential strain in the case of inclined cracks, which could be used to calculate the crack width by multiplying it by the spacing between two cracks. EC2 equation [7.9] is not applicable for mean differential strain, as the stress $\sigma_s$ in the reinforcement bar is based on the assumption that the bar is orthogonal to the crack, which is not the case here. However, as mentioned in section 5.1 of these recommendations, MC2010 does give an indication of how to perform this calculation, although the equation given appears to be inaccurate as no indication is given for the method of calculating the mean total strain $\varepsilon_{\perp}$. For concrete ties, the mean total strain is the mean strain of the reinforcement bar, which is clearly given in MC2010 equation [7.6.5]. This equation could also be used in this case, but not without a number of concerns as described below.

In practice, the designer will have access to the results of a structural analysis giving the strain tensor $\overline{\overline{\varepsilon}}$ and the membrane stress tensor $\overline{\overline{N}}$ at each point, or in each finite element.

The method proposed in section 5.5 is based on the assumption that if the stress distribution given by the structural analysis is acceptable, then accounting locally for cracking does not affect the equilibrium. This approach enables the forces in the reinforcement bars and concrete in the strut and tie model to be calculated and may be divided into the following three steps:

1) For the combination of loads that create the crack, the crack angle $\theta$ is calculated from the structural analysis as described in section 5.2. Once the crack has formed, this angle is fixed for all subsequent combinations of loads.

The membrane forces $N_r$, $T_r$, and $N_{//}$ per unit length applied in the direction of the crack are calculated from the membrane stress tensor determined from the structural analysis, and defined in the reference space O$x$, O$y$ by:

$$\overline{\overline{N}} = \begin{bmatrix} N_{xx} & N_{xy} \\ N_{yx} & N_{yy} \end{bmatrix}$$

As the reference space ($t$, $n$) makes an angle of - $\theta$ with the O$x$ axis (see Figure 5.3), the membrane forces $N_r$ and $T_r$ are calculated according to the projections on the reference axes ($t$,$n$) by:

$$N_r = \vec{n}.\overline{\overline{N}}.\vec{n}$$
$$T_r = \vec{t}.\overline{\overline{N}}.\vec{n}$$

The force $N_{//}$ is the force normal to a plane perpendicular to the crack.

From which can be derived the following equations:

$$N_{//} = \vec{t}.\overline{\overline{N}}.\vec{t} \qquad [5.3]$$

– $N_r = N_{xx}\sin^2\theta + N_{yy}\cos^2\theta - 2N_{xy}\sin\theta\cos\theta$ force normal to the crack;

– $T_r = (N_{yy} - N_{xx})\sin\theta\cos\theta + N_{xy}(\cos^2\theta - \sin^2\theta)$ force tangential to the crack, zero on creation of the crack;

– $N_{//} = N_{xx}\cos^2\theta + N_{yy}\sin^2\theta + 2N_{xy}\sin\theta\cos\theta$ compression of the strut parallel to the crack.

NOTE.– For the combination of loads that creates the crack at a given point, the crack is orthogonal to the direction of the main tensile stress (see section 5.2). Therefore, $T_r = 0$.

The corresponding angle $\theta_r$ is given by the equation:

$$\operatorname{tg} 2\theta_r = 2\frac{N_{xy}}{N_{xx}-N_{yy}} \qquad [5.4]$$

2) The opening of the crack is not unconstrained, and several phenomena oppose its opening, starting with the tension in the reinforcement bars. Other factors include the interlocking of the aggregates in the concrete, and the dowel resistance of the reinforcement bars which oppose the relative tangential slippage $u$ of the two crack edges.

For simplification, the approach described below only considers the stresses provided by tension in the reinforcement bars ("free" crack). Similarly, this approach is limited to the case of crack creation when $T_r$ is zero (however, see section 5.9.1. Example 1 where $T_r$ is different from 0). The addition of the other terms of constraints does not result in any particular difficulty, except perhaps for the non-linearity of the equations.

In order to maintain the equilibrium provided by the membrane forces in the structural analysis, $N_r$ must be balanced throughout the length of the crack as it is being formed. The balancing forces are the tensions in the reinforcement bars, i.e. $F_{sx}$ and $F_{sy}$, the forces acting on the reinforcement bars along a unit length of the crack (see Figure 5.3),

$$\begin{cases} F_{sx} = N_r \sin\theta \\ F_{sy} = N_r \cos\theta \end{cases} \qquad [5.5]$$

3) The equilibrium at the crack is replaced by a "mean" equilibrium between two cracks (in stabilized cracking). This enables the mean strains to be obtained. Between the cracks, part of the tension is absorbed by the concrete in the direction perpendicular to the crack, and this is therefore subtracted from $N_r$ (see Figure 5.4). The mean value of this tension is subtracted from the calculated tension in the reinforcement bars at the crack, and the mean stress in the reinforcement bars and hence their mean strain (tension stiffening) can then be calculated.

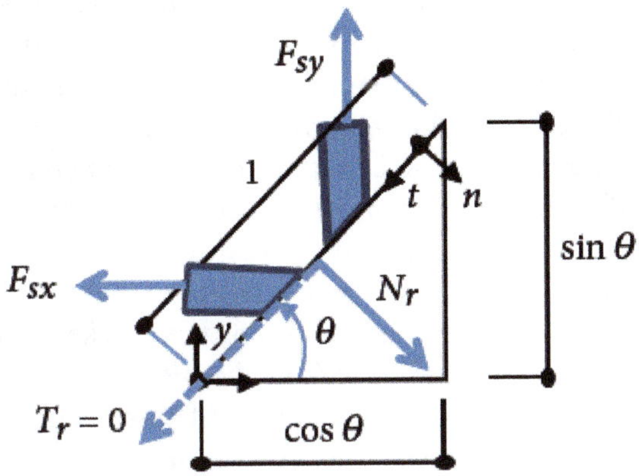


**Figure 5.3.** *Balance of forces along the crack*

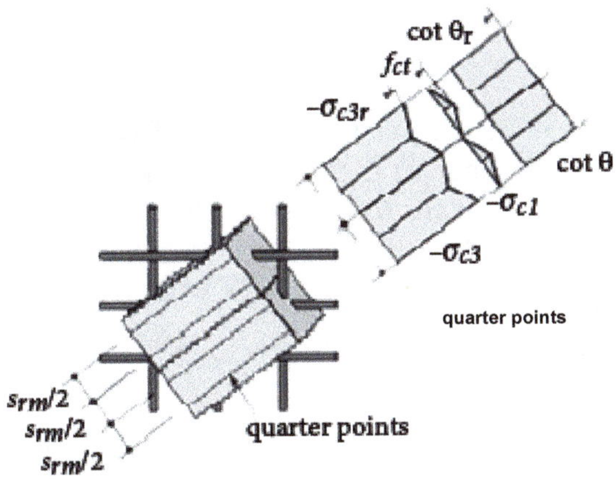


**Figure 5.4.** *Variation in the tensile and compressive stresses in the concrete in a strut-after [KAU 98]*

The inclined spacing between two cracks (or the virtual anchorage length as given in MC2010), $s_r$, is determined by the spacing of two virtual ties enclosing the horizontal and vertical reinforcement bars. The thickness of the virtual ties is given by:

$$h_{eff} = \min\,(h/2\,;\; 2.5(c + \phi/2))$$

On each face of the wall, where $h$ is the total thickness. For simplicity, this effective thickness is assumed to be identical on the two faces. Between the cracks, and perpendicular to them, the tensile stress in the concrete varies between zero at the crack and a maximum of $f_{ct}$ as shown in Figure 5.4. By convention, this stress is applied to the layer of concrete of thickness $h_{eff}$ covering the reinforcement bars on each face of the wall. The mean tension transmitted to the concrete is therefore equal to $2\beta f_{ct} h_{eff}$ taking a mean value for the effective thickness. If the tension in the concrete varies linearly, and its maximum value of $f_{ct}$ is reached between two cracks, then $\beta$ will be equal to 0.5. The codes give a value of 0.6 in the situation under consideration. The case to be considered in the remainder of this section is that when the loads under consideration create the crack. In this case, $T_r = 0$, and this determines $\theta$. In this case, the mean tensions in the reinforcement bars are given by:

$$\begin{cases} F_{sx} = (N_r - 2\beta f_{ct} h_{eff})\sin\theta \\ F_{sy} = (N_r - 2\beta f_{ct} h_{eff})\cos\theta \end{cases} \qquad [5.6]$$

From Figure 5.3, it can be seen that the section of vertical reinforcement bars (or horizontal, case dependent) is that passing through a horizontal (or vertical) length of wall equal to cos $\theta$ (or sin $\theta$). Hence, the mean stresses in the reinforcement bars can be derived:

$$\sigma_{sx} = \frac{N_r - 2\beta f_{ct} h_{eff}}{2 h_{eff} \rho_{sx,eff}}$$

$$\sigma_{sy} = \frac{N_r - 2\beta f_{ct} h_{eff}}{2 h_{eff} \rho_{sy,eff}}$$

The strains in the reinforcement bars can then be obtained by dividing by Young's Modulus for the steel used:

$$\varepsilon_{sx} = \frac{1}{E_s}\left(\frac{N_r - 2\beta f_{ct} h_{eff}}{2 h_{eff} \rho_{sx,eff}}\right) \qquad [5.7]$$

$$\varepsilon_{sy} = \frac{1}{E_s}\left(\frac{N_r - 2\beta f_{ct} h_{eff}}{2 h_{eff} \rho_{sy,eff}}\right)$$

In this expression:

– $\beta$ (or $k_t$, in EC2) is the coefficient representing the mean value of the tension in the concrete once the cracking has stabilized. This coefficient is usually given as 0.6 in EC2 and MC2010; however it has been shown that this value is too high, and the tension stiffening $\frac{\beta\sigma_{sr}}{E_s}$ should be modified by multiplying $\beta$ by a coefficient of 0.6 (see section 3.4);

– $h_{eff}$ is the effective thickness on each face of the wall, and is taken to be equal to:

$$h_{eff} = \min\,(h/2;\ 2.5\ (c + \phi\ /\ 2))$$

in the codes, which has an impact on thick walls. In the case of thin walls, this expression reduces to *h/2* (i.e. half of the total thickness);

– $\rho_{sx,eff}$ and $\rho_{sy,eff}$ are the reinforcement ratios in the two directions, relative to the effective thickness.

## 5.6. Calculating the crack width from reinforcing bar strains

In section 5.5, only the strains in the horizontal reinforcement bars $\varepsilon_{s,x}$ and vertical reinforcement bars $\varepsilon_{s,y}$ were calculated. These are the mean strains, either between two cracks or over a length equal to 2 $l_{s,max,\theta}$. Any distortion $\varepsilon_{s,xy}$ is not significant as the strains in the reinforcement bars are linear.

By analogy with concrete ties, it would appear relevant to assume that one could make the mean total strains equal to those of the reinforcement bars in the x and y directions, that is

$$\varepsilon_{xx} = \varepsilon_{s,x} \text{ and } \varepsilon_{yy} = \varepsilon_{s,y}$$

However, the mean total distortion $\varepsilon_{xy}$ has not yet been determined. Between cracks in the interior of the strut, this is not zero and, although low, it may be calculated. Most of the distortion is due to the crack width, as shown below.

Assuming that the distortion can be negligible, the resulting mean total strain in the direction perpendicular to the crack is derived from the usual base change formulae.

The mean strain of the concrete is due partly to the tension in the concrete within the thickness of the strut (see Figure 7.6.6 of MC2010), and is therefore orthogonal to the crack, and partly to the Poisson effect from the compression of the strut. Hence:

$$\varepsilon_{\perp} - \varepsilon_{c,\perp} = \varepsilon_{s,x} \sin^2\theta + \varepsilon_{s,y} \cos^2\theta - \varepsilon_{c,\perp}$$

$$= \frac{1}{E_s}\left[\left(\frac{N_r}{2h_{eff}} - \beta f_{ctm}\right)\left(\frac{\sin^2\theta}{\rho_{sx,eff}} + \frac{\cos^2\theta}{\rho_{sy,eff}}\right) - \beta\alpha_e f_{ctm} + \nu\frac{\alpha_e N_{//}}{h}\right] \quad [5.8]$$

where

$$\alpha_e = \frac{E_s}{E_c}.$$

The final term due to the Poisson effect is small and can be neglected.

NOTE.–

– in reality, the strains in the reinforcement bars do not allow the establishment of a plane tensor, and the above approach does not take account of the distortion occurring at the crack. Comparison with CEOS.fr experimental results indicates that this could result in the crack width being underestimated. However, the result is satisfactory if within an order of magnitude;

– this equation should not be applied close to the limiting angles of 0 and 90°, partly because of the discrete distribution of the reinforcement bars, and partly because the components $s_{rx}$ and $s_{ry}$ of $s_r$ can no longer be the crack spacing in the two virtual horizontal and vertical ties as the strut is too inclined. In this case, the strut-and-tie model itself is no longer relevant;

– care should be taken to use the mean cover in the calculations;

– the value of $\beta$ may be taken as that given in MC2010, depending on the type of loading. However, a lower value is preferable, as shown in the analysis of concrete ties. It would be possible to take account of the effect of confinement of the reinforcement bars, resulting in different values of $\beta$ in the horizontal and vertical directions, but this effect would certainly be more sensitive to the spacing than to the tension stiffening.

This expression takes no account of a number of favorable factors, including interlocking, dowel effect and improved bond strength due to compression in the strut.

As the mean total strain is then assumed to be equal to the mean strain of the reinforcement bars, the crack width $w_d$ can be calculated in accordance with MC2010:

$$w_d = 2 \cdot l_{s,max} \left( \varepsilon \perp - \varepsilon c, \perp \right)$$

In conclusion, this approach of equating the mean total strain in the x and y directions with the strains in the reinforcement bars is not exact. It can be shown that these quantities are different. The total mean distortion must also be taken into account. [5.8] may be used to calculate the crack width. However, an alternative and more appropriate approach to the calculation of the crack width is given in section 5.7.

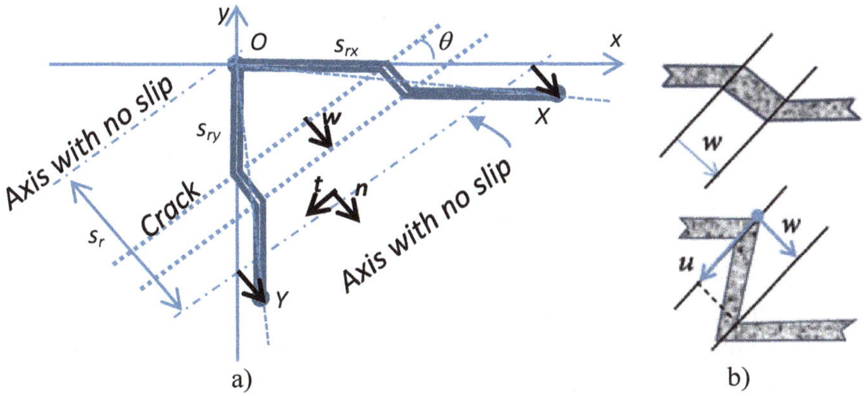


**Figure 5.5.** *a) Deformation of the reinforcement bars; b) distortion of reinforcement bars*

## 5.7. Calculating the crack width in accordance with the strut and tie model

As described in the conclusion of section 5.6, the approach based on [5.8] is not entirely representative of the physical model on which it is based. According to the model, there is no slippage between the reinforcement bars and the concrete along the median axes between the cracks (symmetry in stabilized cracking) or, if the cracking is not stabilized, at a distance of

$l_{s,max,\theta}$ on both sides of the crack. That is to say that the extension of the reinforcement bar between the points O and $M_x$ in Figure 5.5(a) (or O and $M_y$) is equal to the extension of the concrete plus the crack width in the direction under consideration. If the tangential displacement is zero, each reinforcement bar will stretch by $w$ (see Figure 5.5(b)). If it is not zero, the extension will be based on the expression $\sqrt{u^2 + w^2}$, which introduces a non-linearity. The behavioral model will need to be extended to take account of the effects that have not been considered here. However, if $u$ is small compared with $w$, its effect is negligible, enabling the linearity of the equations to be retained. This approximation is used in section 5.7.

Hence, in this case, two kinematic compatibility conditions can be derived, expressing the extensions in the $x$ and $y$ reinforcement bars respectively, and maintaining the fixed bond between the reinforcement bars and the concrete at the points O, $M_x$ and $M_y$:

$$(\varepsilon_{c//} \cos^2\theta + \varepsilon_{c\perp} \sin^2\theta) s_{rx} + \frac{w}{\sin\theta} - u \cos\theta = \varepsilon_{s,x} s_{rx}$$

$$(\varepsilon_{c//} \sin^2\theta + \varepsilon_{c\perp} \cos^2\theta) s_{ry} + \frac{w}{\cos\theta} + u \sin\theta = \varepsilon_{s,y} s_{ry}$$

These two equations are compatible when the reinforcement bars are dimensioned in accordance with [5.1]. If this is not the case, the full (non-linear) equations must be solved, taking account of the extension $\sqrt{u^2+w^2}$ and possibly also the effects of interlocking and dowel effect. If the compatibility conditions are satisfied, this leads to the following linear approximation for the crack width:

$$w = 2l_{s,max}\left[\varepsilon_{s,x}\sin\theta + \varepsilon_{s,y}\cos\theta - \varepsilon_{c//}(\cos\theta + \cos\theta)\sin\theta\cos\theta - \varepsilon_{c\perp}(\sin^3\theta + \cos^3\theta)\right]$$

$$w = \frac{2l_{s,max,\theta}}{E_s} \cdot \left[\begin{array}{c} \left(\frac{N_r}{2h_{eff}} - \beta f_{cm}\right)\left(\frac{\sin\theta}{\rho_{sx,eff}} + \frac{\cos\theta}{\rho_{sy,ff}}\right) \\ -\alpha_e\left[\left(-\nu\beta f_{ct} + \frac{N_{//}}{h}\right)(\cos\theta + \sin\theta)\sin\theta\cos\theta + \left(\beta f_{ct} - \nu\frac{N_{//}}{h}\right)(\sin^3\theta + \cos^3\theta)\right] \end{array}\right] \quad [5.9]$$

In this expression, the influence of the strain in the concrete is low and may be neglected. The equation gives a conservative evaluation, particularly as it ignores both interlocking and dowel effect.

Finally, the compatibility of the strain in the concrete between the points $M_x$ and $M_y$ should be considered. If the strain in the concrete is neglected, this distance does not vary as the crack opens. This suggests that distortion (i.e. a variation in the mean angle between the two reinforcement bars on and around the crack, see Figure 5.5(a)) should be considered. This distortion is directly related to the crack width, introducing the distortion due to the concrete remaining in its elastic domain:

$$\alpha = 2\varepsilon_{xy} = \left[\frac{w}{l_{s,max,\theta}} - \frac{1}{hE_c}(N_{//} - 2\beta f_{ct} h_{eff})\right] sin\theta cos\theta \qquad [5.10]$$

The compression of the strut and the transverse tension increase the amplitude of this distortion.

This expression explains the origin of the distortion in the wall, that is the relative horizontal displacement between the top and bottom of the wall. This relationship between the quantities in the two sides of the equation has been verified when the cracking is stabilized and the post-elastic damage to the concrete between cracks is not too severe. If this is not the case, additional distortion will be caused.

## 5.8. Recommendations for evaluating the cracking in walls subject to earthquake situations

A seismic situation is usually considered to be a quasi-accidental situation, with earthquake resistance verified at ULS. Current construction standards do not require the cracking to be verified in this situation. However, some special works may require the water or air-tightness, or containment function of the wall to be verified following an earthquake. This raises the question of residual crack width, which is that remaining following the earthquake. From this point of view, the maximum crack width during the earthquake is not relevant, although it could be if equipment is anchored to the wall and its stability or functionality is required in an earthquake situation. Finally, during an earthquake, the stress is cyclic and reversing, creating two networks of cracks that are relatively symmetrical about the vertical axis. This results in more damage to the concrete than a simple monotonic stress.

The following conclusions may be drawn from test results:

– in a ULS situation, the contribution of the concrete to the tensile strength is less than that at SLS and the coefficient $\beta$ should be multiplied by a factor of around 2/3 (two third);

– the earthquake induced cyclic alternating stress causes more damage due to the dual network of cracks, and thus the maximum strength and the cracked stiffness are both reduced by around 10 %;

– the maximum crack width during the earthquake increases at a faster rate than the simple proportionality to the stress given in the codes after the initial crack formation. The crack width can be evaluated using the equations suggested in sections 5.4 and 5.7 as long as the yielding of the reinforcement bars remains limited. However, beyond yielding, the models underestimate the crack width;

– the residual crack width remaining after the end of the earthquake forms soon after the crack opens, mainly due to interlocking effects, and may reach up to half the maximum crack width under certain conditions. However, as long as the reinforcement bars do not significantly exceed their elastic limit, as is the case at ULS when yielding remains localized, the residual crack width does not depend significantly on the maximum crack width attained, and crack width values observed rarely exceed 0.1 mm.

Experimental results given in the main deliverable LP3, *Results obtained in the understanding of cracking phenomena*, (PN CEOS.fr; LP3 -, 2013), provide further justification of the above recommendations.

## 5.9. Examples of application of cracking equations in a wall subjected to a shear stress in the plane of the wall

The following examples represent the experimental conditions applied to CEOS.fr project wall No. 3, subjected to a force pushing in a single direction. The calculation methods applied are derived from the equations given in *fib* MC2010.

The dimensions of the wall are 4200 mm (length) × 1050 mm (height) × 150 mm (thickness).

The concrete is Class C40/50, with a mean tensile strength of 3.5 MPa and an instantaneous modulus of 36.3 GPa.

Given the specific boundary conditions applying to the upper and lower load-spreading beams, and the interaction of the wall with the metal framework enclosing it, a detailed finite element model of all the structures and the internal and external interfaces between them is used to model the behavior of the wall. This model is shown in Figure 5.6 (excludes the steel framework).

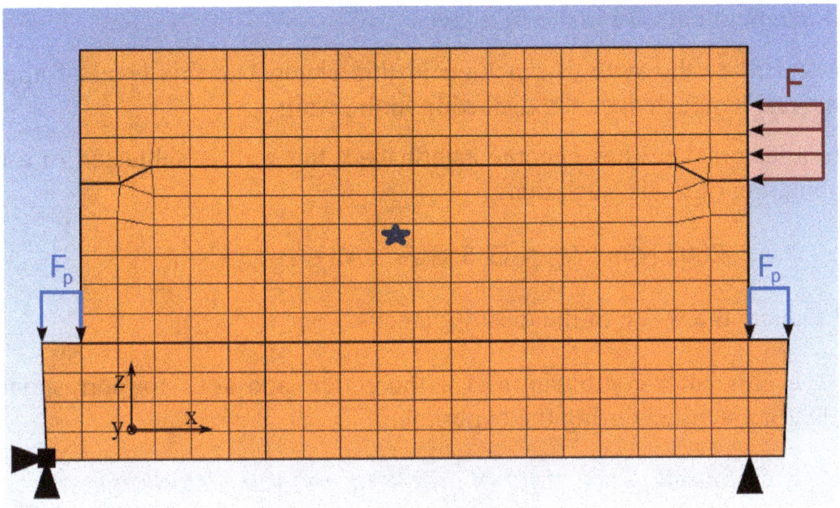


**Figure 5.6.** *Wall model subjected to a shear stress (crack analysis location marked with a star)*

The behavior modeled is perfectly plane in nature; therefore the results of the analysis are membrane forces. A number of constitutive laws of reinforced concrete were tested. These can give varying results locally, even when the overall behavior in terms of force and displacement is correctly represented.

The model suggested here uses the results of a traditional elastic analysis, as this is the constitutive model most often used in practice.

Under the effect of a force of 2.4 MN, the membrane stresses obtained in the element * pointed on Figure 5.6 are as follows ($x$ = horizontal, $y$ = vertical):

$$N_{xx} = -399 \text{ kN/m} ; N_{yy} = 13.8 \text{ kN/m} ; N_{xy} = -485 \text{ kN/m}$$

The assumptions described above apply to all three of the following application examples.

### 5.9.1. *Example 1*

The reinforcement of the wall consists of a grid of reinforcement bars on each face, with an outer cover of 10 mm. The horizontal and vertical reinforcements are identical, i.e.10 Ø 10 mm per meter, and the horizontal reinforcement bars are in the first layer:

a) calculate the state of cracking in this element at this level of applied stress, and hence deduce the distortion in the wall;

b) after cracking has become established, the wall is subjected to a new combination of loads, such that:

$N_{xx}$ = -200 kN/m ; $N_{yy}$ = 13.8 kN/m ; $N_{xy}$= - 485 kN/m

Evaluate the width of the existing cracks.

NOTE.– This latter combination B is theoretical and does not correspond to any loading applied during the experiment.

#### 5.9.1.1. *Calculating the state of cracking and the distortion in the wall*

a) The first step is to determine the principal directions and the principal membrane stresses. As the quantities are determined from the diagonalization equations, this calculation can easily be carried out using a calculator or spread sheet. Most finite element software packages give these results directly.

MC2010 indicates that the direction of the crack is orthogonal to the direction of the main tensile stress. This approximation is sufficiently accurate as the results are only slightly sensitive to a limited variation in the angle given the dispersion of the cracking phenomenon. Applying the usual equations for the principal directions of the membrane stress tensor and the principal associated stresses gives:

$$\theta_r = \frac{1}{2} arctg\left(2\left|\frac{N_{xy}}{N_{xx}-N_{yy}}\right|\right) = 33.5°$$

Crack angle / x: see equation [5.4]

– Principal compressive stress:

$$N_{//} = \frac{1}{2}\left(N_{xx} + N_{yy} - \sqrt{\left(N_{xx} - N_{yy}\right)^2 + 4N_{xy}^2}\right) = -719\ kN/m$$

– Principal tensile stress:

$$N_r = \frac{1}{2}\left(N_{xx} + N_{yy} + \sqrt{\left(N_{xx} - N_{yy}\right)^2 + 4N_{xy}^2}\right) = 334\ kN/m$$

NOTE.–

The principal stresses may also be calculated from [5.3], using a value of $\theta_r = 33.5°$, and may be independently verified by the reader.

MC2010 indicates that the crack must be inclined sufficiently for the remainder of the procedure to be applicable. In practice, this is the case if the angle is greater than 15° to the horizontal and vertical axes, as it is here.

b) The second step is to evaluate the crack spacing.The spacing to be used is given by the Vecchio and Collins equation (see section 5.4):

$$S_{rm} = \left(\frac{\sin\theta_r}{S_{rmx}} + \frac{\cos\theta_r}{S_{rmy}}\right)^{-1}$$

where the components $s_{rmx}$ and $s_{rmy}$ are the crack spacing in the two virtual ties, one in the horizontal direction, the other in the vertical direction, taking account of the reinforcement bars in each direction. In MC2010, these virtual crack spacing are replaced by the anchorage length $l_{s,max}$ of the reinforcement bars in each of the two directions, as given by the equation:

$$l_{s,max} = kc + \frac{1}{4}\frac{f_{ctm}}{\tau_{bms}}\frac{\phi_s}{\rho_{s,eff}}$$

where $c$ is the cover. As there are two layers of reinforcement bars on each face of the wall:

– the first option is to take the cover for each layer and calculate the anchorage lengths for each separately;

– the second option is to use the mean cover of the two layers, in this case 15 mm.

Either option is acceptable (see the table below):

– $k$ is a factor which may be taken as being equal to 1,

– $\frac{f_{ctm}}{\tau_{bm}}$ is the ratio between the mean tensile strength of the concrete and the mean bond strength between the reinforcement bar and the concrete, assuming this to be constant along the anchorage length. According to MC2010, this ratio is independent of $f_{ctm}$, depending only on the loading conditions. In the case of short term loading (as in CEOS.fr experiment), this ratio is equal to 1/1.8;

– $\phi_s$ is the diameter of the reinforcement bars in the direction under consideration, in this case 10 mm in both directions;

– $\rho_{s,eff}$ is the effective percentage of the reinforcement bars in the direction under consideration. In the case of a wall, it is calculated from the effective thickness of the concrete on each face of the wall, as given by:

$$h_{eff} = \min\,(h/2;\ 2.5\,(c + \phi/2))$$

In the case of a thick wall, as in this example, the effective thickness is determined by the second term. There is therefore a region of 'neutralized' concrete in the centre of the wall thickness that makes no contribution to the tensile strength. As in the previous case, it is possible to take either two different values for the cover, leading to two different reinforcement percentages in the two directions, or to take the mean cover. In the second case, the effective thickness is given by:

$2.5 \times \left(15 + \frac{10}{2}\right) = 50\ mm$), on each face.

The effective percentage is then the same in each of the two directions and equal to:

$$\frac{10 \times \pi \times (5 \times 10^{-3})^2 m^2}{0.050 \times 1 m^2} = 1.57\%$$

This is the usual value, given in MC2010, that is the cross-sectional area of the reinforcement bars as a proportion of the total cross-sectional area.

The following table summarizes the calculations using each of the options relating to the cover. When using the mean cover, the anchorage length is the same in both directions, as the reinforcement layers are identical.

| | $c$ (mm) | $h_{eff}$ (mm) | $\rho_{s,eff}$ (%) | $l_{s,max} = \frac{s_{rmax,x\ ou\ y}}{2}$ (mm) | $s_{r,max}$ $(2l_{s,max,\theta})$ (mm) |
|---|---|---|---|---|---|
| Mean cover | $c_m = 15$ | 50.0 | 1.57 | 103 | 149 |
| Separate cover | $c_x = 10$ | 37.5 | 2.09 | 76 | 147 |
| | $c_y = 20$ | 62.5 | 1.26 | 131 | |

**Table 5.1.** *Crack spacing calculated according to the concrete cover*

The difference between the two results is sufficiently small for either result to be acceptable. It should be noted that MC2010 gives a crack spacing higher than that found in experimental results. However, this crack spacing is still more closely correlated to the experimental results than that given by EC2-1-1.

NOTE.– The values of $s_r$ given in the table are maximum values. The mean values are obtained by dividing by 1.7.

c) The third step is to calculate the stresses in the reinforcement bars, then use these to calculate the strains (see section 5.5). The tensions in the two reinforcement layers are calculated from the equilibrium at the crack as the membrane forces in the wall are transmitted by only the reinforcement bars alone. As the tangential force is assumed to be zero, we have:

$$\begin{cases} F_{sx} = N_r \sin\theta_r = 334\ \text{kN/m} \times \sin(33.5°) = 184\ \text{kN/m} \\ F_{sy} = N_r \cos\theta_r = 334\ \text{kN/m} \times \cos(33.5°) = 279\ \text{kN/m} \end{cases}$$

$N_r$ is applied over a unit length of crack (1 meter). The cross-sectional area of the horizontal (or vertical) reinforcement bars crossing in this length is 20 × (cross-sectional area of $\phi$ 10 mm) × sin $\theta_r$(or cos $\theta_r$) = 8.66 cm$^2$ (or 13.10 cm$^2$).This can then be used to calculate the stresses in the reinforcement bars at the crack, followed by their maximum strains using a Young's Modulus of 200 GPa:

$$\sigma_{s,x} = \frac{184 \times 10^{-3}\ \text{MN/m}}{8.66 \times 10^{-4}\text{m}^2} = 213\ MPa \Rightarrow \varepsilon_{s,x} = \frac{213\ \text{MPa}}{200\ 000\ \text{MPa}} = 1.06 \times 10^{-3}$$

$$\sigma_{s,y} = \frac{279 \times 10^{-3}\ \text{MN/m}}{13.10 \times 10^{-4}\text{m}^2} = 213\ MPa \Rightarrow \varepsilon_{s,y} = \frac{213\ \text{MPa}}{200\ 000\ \text{MPa}} = 1.06 \times 10^{-3}$$

As the two reinforcement layers are identical, $\theta_r$ does not appear explicitly in the values of the stresses and strains, as can be seen clearly in the equations. The result obtained is the strain in the reinforcement bars at the crack. The reinforcement bars are therefore well within the elastic domain. The mean strains are obtained by reducing the tension in the reinforcement bars via the contribution of the concrete in absorbing the tension between the cracks. Recognizing the contribution of the concrete is directly linked to the variation in the tensile stress within the strut, perpendicularly to the crack; it can be considered that the concrete contribution directly reduces $N_r$, which is applied in the same direction. This contribution is associated with the bond between the reinforcement bars and the concrete, and therefore only applies to the effective thickness.

The contribution of concrete is calculated in accordance with MC2010, taking the mean value for the cover. It is given as a function of the tensile strength of the concrete by the expression:

$$N_{cm} = \beta f_{ct} \times 2h_{eff} \times (1 + \rho_{s,eff}\alpha_e)$$

$$N_{cm} = 0.6 \times 3.5 \times 10^6 \text{ MPa} \times 2 \times 0.050 \text{ m} \times \left(1 + 0.0517 \times \frac{200}{36.3}\right) = 0.228 \text{ MN/m}$$

where $\beta$ is the coefficient expressing the mean contribution of the concrete. With the loading conditions under consideration here, $\beta$ is equal to 0.6 (see MC2010, Table 7.6.2). It should be noted that there is only one effective thickness in this calculation model, and therefore only one cover. It is therefore necessary to use, consistently through the calculation, either the mean cover from the beginning of the procedure, or the mean of the effective thicknesses together with the mean of the percentages associated with the two covers.

Hence, replacing $N_r$ = 334 kN/m by $N_r - N_{cm}$ = 106 kN/m, the mean strains in the reinforcement bars are given by:

$$\varepsilon_{sm,x} = \varepsilon_{sm,y} = 0.339 \times 10^{-3}$$

The final step is to calculate the crack width. This is given by the expression:

$$w_d = 2l_{s,max}[\varepsilon_{s,x} sin\theta + \varepsilon_{s,y} \cos\theta - \varepsilon_{c//}(cos\theta + sin\theta) sin\theta cos\theta - \varepsilon_{c\perp}(sin^3\theta + cos^3\theta]$$

For concrete, the strains are derived from the stresses, which in turn are derived from the membrane forces acting across the entire thickness of the wall. The Poisson coefficient for concrete is taken to be equal to 0.2, as it is assumed that the concrete is not cracked inside the strut.

$$\varepsilon_{c\perp} = \frac{1}{E_c}(N_r - \nu N_{//})$$

$$\varepsilon_{c\perp} = \frac{1}{36\,300\ MPa \times 0.15\ m}(0.334\ MPa + 0.2 \times 0.719\ MPa) = 0.878 \times 10^{-4}$$

$$\varepsilon_{c//} = \frac{1}{E_c}(N_{//} - \nu N_r)$$

$$\varepsilon_{c//} = \frac{1}{36\,300\ MPa \times 0.15\ m}(-0.719\ MPa - 0.2 \times 0.334\ MPa) = -1.443 \times 10^{-4}$$

In the expression for $w$, the strain due to the concrete is therefore equal to $0.264 \times 10^{-4}$, with the strain due to the reinforcement equal to $0.339 \times 10^{-3} \times (\sin(33.5°) + \cos(33.5°)) = 4.70 \times 10^{-3}$. Hence, the influence of the strain in the concrete is therefore negligible compared to the reinforcement, given the presence of the sine and cosine raised to the third power. The strain in the reinforcement bars is the dominant value.

Hence the calculated crack width, taking account of the calculated anchorage length:

$$w_d = 149\ \text{mm} \times 0.470 \times 10^{-3} = 70\ \mu\text{m}$$

The high conventional value of $f_{ctm}$ used in the expression for concrete contribution should be noted. In the absence of a more accurate analysis, which in particular takes account of the scale effect (see section 2.2.1), an equivalent method of accounting for this phenomenon is to multiply the value of $\beta$ by a coefficient of 0.6. This approach reduces the value of $\beta$ to $0.6 \times 0.6$, i.e. $\beta = 0.36$ (see section 3.4), enabling an improved correlation with the experimental results obtained from the CEOS.fr project. Using this lower value reduces the contribution of the concrete to $N_{cm} = 133$ kN/m, thereby increasing the stress and the mean strain in the reinforcement bars. Repeating the calculation gives a crack width value of $w_d = 132\ \mu\text{m}$.

Once the crack width has been determined in this way, the local distortion due to the opening of this crack can be calculated using the following expressions:

$$\alpha = -\left[\frac{w}{l_{s,max,\theta}} - \frac{1}{hE_c}\left(N_{//} - 2\beta f_{ct}h_{eff}\right)\right]\sin\theta\cos\theta$$

$$\alpha = -\left[\frac{2\times 0.070}{149} - \frac{1}{0.15\times 36\,300}(-0.719 - 0.228)\sin(33.5°)\cos(33.5°)\right]$$

$$\alpha = -5.13\times 10^{-4}$$

In this case, the contribution from the concrete strut distortion is not negligible. This expression assumes that the cracking is stabilized. The overall distortion in the wall, that is the ratio between the relative horizontal displacement and the height, results from an approximation sort of the mean of the distortions along the length of the wall. At any given abscissa, the relative displacement may be calculated by integrating the local distortions given along the same vertical axis in the finite element model. In this example, if the distortion of the element under consideration is assumed to be homogeneous along the vertical axis, then the relative horizontal displacement between the top and the bottom is given by:

$$\delta = 0.513 \times 10^{-3} \times 1.05 \text{ m} = 0.54 \text{ mm}$$

### 5.9.1.2. *Evaluating the widths of existing cracks*

A new combination of loads (theoretical in this case) is applied after cracking has become established. This results in the following membrane forces:

$$N_{xx} = -200 \text{ kN/m};\ N_{yy} = 13.8 \text{ kN/m};\ N_{xy} = -485 \text{ kN/m}$$

$N_{xx}$ is the only value that has changed since the previous example, although now the crack shape is known and the angle $\theta_r$ is therefore fixed. The crack no longer corresponds to the principal direction of the stresses, and there is therefore a force tangential to the crack. This force, per unit length of the crack, is given by:

$$T_r = \left(N_{yy} - N_{xx}\right)sin\theta_r cos\theta_r + N_{xy}(cos^2\theta_r - sin^2\theta_r)$$

$$T_r = (13.8 + 200)\ \text{kN/m} \times \sin(33.5°)\cos(33.5°) - 485\ kN/m \times [cos^2(33.5°) - sin^2(33.5°)]$$

$$T_r = 91.4\ kN/m$$

The other forces must also be recalculated:

$$N_r = N_{xx}sin^2\theta + N_{yy}cos^2\theta - 2N_{xy}sin\theta cos\theta$$

$$N_r = -200\ kN/m \times sin^2(33.5°) + 13.8kN/m \times cos^2(33.5°) + 2 \times 485\ kN/m \times \sin(33.5°)\cos(33.5°)$$

$$N_r = 395\ kN/m$$

$$N_{//} = N_{xx}cos^2\theta + N_{yy}sin^2\theta + 2N_{xy}sin\theta cos\theta$$

$$N_{//} = -200\ kN/m \times cos^2(33.5°) + 13.8kN/m \times sin^2(33.5°) - 2 \times 485\ kN/m \times \sin(33.5°)\cos(33.5°)$$

$$N_{//} = -581\ kN/m$$

The calculation model is still valid, as there is tension perpendicular to the crack and compression of the strut. The value of $N_{cm}$ is unchanged.

In performing the calculation, the main difference from case A is the equilibrium at the crack, as account must now be taken of the tangential force. This leads to a new expression for the tensions in the reinforcement bars at the crack:

$$F_{s,x} = (N_r - N_{cm})sin\theta_r + T_r cos\theta_r$$

$$F_{s,x} = (395 - 228)kN/m \times \sin(33.5°) - 91.4\ kN/m \times \cos(33.5°) = 15.6\ \text{kN}/m$$

$$F_{s,y} = (N_r - N_{cm})cos\theta_r - T_r sin\theta_r$$

$$F_{s,y} = (395 - 228)\ kN/m \times \cos(33.5°) + 91.4kN/m \times \sin(33.5°) = 189\text{kN}/m$$

The model is acceptable as long as the reinforcement bars remain in tension.

The remainder of the calculation is similar to that given in the previous example, giving the following results:

$$\sigma_{sm,x}{=}18.03\ MPa \Rightarrow\ \varepsilon_{sm,x}{=}0.090{\times}10^{-3}$$

$$\sigma_{sm,y}{=}144.6\ MPa \Rightarrow\ \varepsilon_{sm,y}{=}0.722{\times}10^{-3}$$

In this case, the tangential force results in the strains being different in the two directions. The strain due to the reinforcement bars is therefore:

$$\varepsilon_s = 0.090 \times 10^{-3} \times \sin(33.5°) + 0.722 \times 10^{-3} \cos(33.5°) = 0.652 \times 10^{-3}$$
$$w = 149\, mm \times 0.652 \times 10^{-3} = 97\mu m$$
$$\alpha = 0.668 \times 10^{-3} \Rightarrow \quad \delta = 0.70\, mm$$

It should be borne in mind that this approach does not take account of any interlocking or dowel effect of the reinforcement bars, which would tend to oppose the tangential force and hence reduce the strains in the reinforcement bars.

### 5.9.2. *Example 2*

Determine the optimum reinforcement under the same applied forces and evaluate the resulting cracking. The state of the cracking is then determined.

The angle of the cracks to the horizontal $\theta_r$ is calculated in Example 1, section 5.9.1. The optimum reinforcement layers dimensions, in terms of cracking (see section 5.4), is such that:

$$\frac{s_{rm,x}}{s_{rm,y}} = cotg\theta_r = \text{cotg}(33.5°) = 1.51$$

which leads to the equation:

$$c + \frac{\emptyset_{s,x}}{2} + \frac{1}{7.2}\frac{\emptyset_{s,x}}{\rho_{s,x,eff}} = 1.51\left(c + \frac{\emptyset_{s,x}}{2} + \frac{1}{7.2}\frac{\emptyset_{s,y}}{\rho_{s,y,eff}}\right)$$

This equation uses the mean cover, equal to the sum of the outer cover $c$ = 10 mm and half the diameter $\phi_{s,x}$ of the horizontal bars. In this case, there are four parameters to be chosen; the diameter and spacing of the bars in each of the two layers.

The equation is simplified if the horizontal (outer) layer is fixed, for example at the previous values of 10 Ø 10 mm per meter. The equation is then reduced to:

$$10 + 5 + \frac{1}{7.2} \times \frac{10}{1.57\%} = 103 = 1.51 \times \left(10 + 5 + \frac{1}{7.2} \times \frac{\emptyset_{s,y}}{\rho_{s,y,eff}}\right) \Rightarrow \frac{\emptyset_{s,y}}{\rho_{s,y,eff}} = 582$$

where $\phi_{s,y}$ is expressed in mm. If this equation is strictly verified, $s_{rm,x} = 206$ mm and $s_{rm,y} = 136$ mm. Then $s_r = 114$ mm and $w = 114 \text{ mm} \times 0.374 \times 10^{-3} = 43$ µm. These values are below those found in the previous example. This is due to the optimization.

All that remains in order to determine the reinforcement requirements is to find the diameter and spacing $e_y$ (mm) of the vertical reinforcement bars. In the following equation $\chi$ is the radius of the reinforcement bar cross-section in mm:

$$h_{y,eff} = 2.5(15+\chi) \Rightarrow \frac{\emptyset_{s,y}}{\rho_{s,y,eff}} = \frac{2\chi}{\frac{\pi \times \chi^2}{2.5(15+\chi) \times e_y}} = 582 \Rightarrow e_y \approx 365 \frac{\chi}{15+\chi}$$

The choice can then be made from the table below, which gives possible solutions of the above equation:

| $\phi_{s,y}$(mm) | $e_y$ (mm) |
|---|---|
| 8 | 77 |
| 10 | 91 |
| 12 | 104 |
| 14 | 116 |

Increasing the diameter of the vertical reinforcement bars from Ø 10 mm to Ø 12 mm results in a lower crack width.

Altering the reinforcement spacing has only a minor effect on the cracking, as it is an optimum (derivative = 0). However, the optimum distribution of the reinforcement bars for strength at ULS may be different.

### 5.9.3. *Example 3*

A wall is subjected to a shear force in the plane of the wall and its reinforcement is known. Evaluate the state of the cracking for a given load, knowing the relative horizontal displacement between the top and bottom of the wall.

On the basis of the data of the CEOS.fr project wall, it is assumed that it undergoes a differential horizontal displacement of 1 mm between the top and bottom of the wall. This corresponds to a distortion of $\frac{1}{1.05\times10^{3}}$ = $0.95 \times 10^{-3}$ which is assumed to be homogeneous.

It is necessary to find the angle of the cracks. This can be done by means of an approximate evaluation of the stresses derived from the vertical load distribution and the shear force, or a better evaluation using finite element modeling. Small variations in the angle have little effect on the evaluation of the cracking. In this case, the angle has already been determined in the first example.

As the crack angle and the arrangement of reinforcement bars are known, the crack spacing is also known. It was determined in Example 1, section 5.9.1: $s_r = 149$ mm.

Ignoring the elastic contribution of the concrete, the distortion gives:

$$\frac{w}{S_r} = \frac{\alpha}{2sin\theta_r cos\theta_r} = \frac{0.95\times10^{-3}}{2\,\sin(33.5°)\cos\,(33.5°)} = 1.03 \times 10^{-3}$$

and hence the crack width:

$$w = 1.03 \times 10^{-3} \times 149 \text{ mm} = 153\ \mu\text{m}$$

This width is only an approximate evaluation as the cracking and distortion are assumed to be homogeneous and the crack angle may be inaccurate. However, this approach does give an indication of the state of the cracking, provided that the concrete is undamaged, which is usually the case at SLS.

# 6

# Minimum Reinforcement of Thick Concrete Elements

The reinforcement of thick elements is characterized by the following factors:

– the behavior of early age concrete due to THM, scale and 3D effects, as these effects are more pronounced in thick elements than they are in thin elements;

– the mechanical behavior when subjected to external loads.

This chapter presents a set of reinforcement rules for early age concrete elements, given that the stabilized cracking level is often not reached at the SLS for these elements.

The calculation of reinforcement needs to be adapted to the function of the element under load. For example, a wall may be subjected either to shear or to flexure depending on the applied load or deformation. In order to avoid the use of unnecessary reinforcement, the role of the reinforcement bars has to be considered in terms of their orientation relative to the direction of the stresses generated by each load or deformation. Tests carried out by CEOS.fr on RL beams (see section 1.2) have shown that surface cracks due to THM effects may be unrelated to cracks occurring under mechanical loading, and the width of these cracks varies little under such loads.

NOTE.– The combination of mechanical effects and deformations is discussed further in Chapter 9.

## 6.1. Reinforcement of reinforced concrete ties

### 6.1.1. *Detailed calculation from a 3D approach*

For thick elements, the three dimensional distribution of strains and stresses when the element is in tension plays a non-negligible role in the occurrence of cracking and reduces the apparent tensile strength of the concrete. This has a direct effect on the cross-sectional area of the reinforcing bars needed to ensure non-fragility requirements in tension.

$$\rho = \frac{A_S}{A_C} \geq \frac{\gamma_0}{\frac{f_{yk}}{f_{ct,scale}} - \alpha_e}$$

$\gamma_0$ is derived from the tensile strength provided by the concrete in a reinforced concrete section before cracking occurs (see section 3.3):

$$F_c = \int_{section} f_{ct,scale}(y,z)dydz = \gamma_0 f_{ct,scale} A_c$$

where $f_{ct,scale}$ is the concrete tensile strength close to the reinforcing bars.

The value of $f_{ct,scale}$ should be calculated taking account of the scale, which depends on the state of cracking of the tie. It corresponds to the minimum value prior to stabilized cracking of the tie under load.

### 6.1.2. *Simplified methodology for calculating concrete reinforcement*

#### 6.1.2.1. *Short-term reinforcement*

The short-term period lasts during the heating and cooling phases of the young age concrete and ends when the self-stresses due to this phenomenon have disappeared.

If no 3D calculation is available, equation [7.1] given in EC2-1 may be used with the coefficients $k_c$ and k derived from measures taken from area in tension of structures in serviceability, in accordance with Table 6.1. Alternatively, the value of the coefficient k may be chosen to be between 0.65 and 1.0 in accordance with EC2-1 section 7.3.2.

| | *Concrete area in tension $A_c$* | *Stress profile* | *Minimum reinforcement and value of $h_t$* |
|---|---|---|---|
| a) | When limited surface cracking of the concrete results from the temperature difference between the core and the surface, the area of the concrete in tension is given by<br>$A_c = 0.20\ h \cdot 1.0\ m^2$<br>with a tensile stress distribution profile in the form of a double triangle | Heating or formwork removal<br>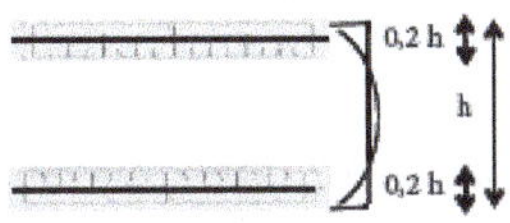 | $A_{smin} = \frac{0.5 \cdot 0.2h \cdot 1 \cdot f_{ctm,scale}}{f_{yk}}$<br>$f_{ctm,scale}$ is calculated for a layer thickness of $h_t$ at maximum stress of:<br>$h_t = 0.2\ h / 3$ |
| b) | When the cracking results from the overall cooling of the element or is due to drying while under restraint, the area of the concrete in tension is given by<br>$A_c = 0.5\ h \cdot 1.0\ m^2$<br>with a quasi-uniform tensile stress distribution profile | Cooling while under end restraint, but no edge restraint<br>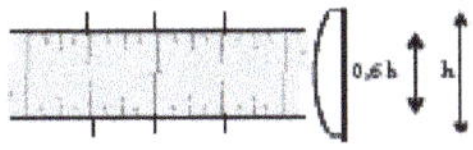 | $A_{smin} = \frac{k \cdot 0.5h \cdot 1 \cdot f_{ctm,scale}}{f_{yk}}$<br>$f_{ctm,scale}$ is calculated for a layer thickness of $h_t$ at maximum stress of:<br>$h_t = 0.6\ h$ |
| c) | When the cracking is due to daily temperature cycle, the thickness $h_t$ is equal to 0.30 m and the area of the concrete in tension is given by<br>$A_c = 0.30 \cdot h\ 1.0\ m^2$<br>with a tensile stress distribution profile in the form of a triangle | Daily temperature cycle<br>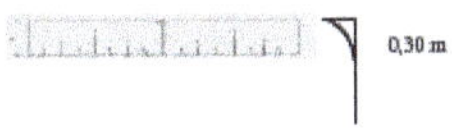 | $A_{smin} = \frac{0.5 \cdot 0.3 \cdot 1 \cdot f_{ctm,scale}}{f_{yk}}$<br>$f_{ctm,scale}$ is calculated for a layer thickness of $h_t$ at maximum stress of:<br>$h_t = 0.3/3 = 0.1$ m |

Notes:

a) In the first case, the tensile zone has a thickness of 0.2 h. This assumes a parabolic variation across the thickness of the wall and a mean stress of zero. It is assumed that the region in maximum tension (close to $f_{ct,m}$), and used to calculate $f_{ctm,scale}$, corresponds to 1/3 of the total tensile zone.

b) In the second case, it is assumed that the cross section in maximum tension (close to $f_{ctm}$), used to calculate $f_{ctm,scale}$, corresponds to $h_t = 0.6$ h. The reinforcement for each surface is half of the total reinforcement, and this is the reason why the coefficient of 0.5 is chosen.

c) In the third case, the tensile zone extends up to a depth of 0.30 m (depth to which the daily temperature cycle propagates) and the zone in maximum tension (close to $f_{ct,m}$), and used to calculate $f_{ctm,scale}$, corresponds to 1/3 of the tensile zone.

These notes also apply to Table 6.3.

**Table 6.1.** *Minimum reinforcement in the short term for concrete thickness $\geq 400$ mm*

Table 6.1 applies the principles for the hydration effect of concrete at an early age as described in Chapter 2. For each situation the table provides:

– a tensile stress profile that can be used to determine the tensile force;

– a thickness $h_t$ of the concrete when subject to high tension, that can be used to reduce the tensile strength according to the scale effect and to calculate $f_{ctm,scale}$.

The first case (Table 6.1(a)) corresponds to the gradient that occurs around 3 days after pouring. In this case, the strength at three days may be calculated taking the scale effect into account, but in this case, the depth $h_t$ of the layer in high tension is relatively less thick.

The second case (Table 6.2(b)) corresponds to a longer period around 10 to 20 days after pouring. In this case, the tensile strength is greater, but the scale effect could be significant as a large proportion of the surface is subjected to high tensile stresses.

The last case (Table 6.1(c)) can occur following the removal of formwork or when the concrete is fully cured. In this case, the scale effect is limited.

The principles outlined in Table 6.1 above may also be adapted to elements whose thickness is not uniform. In this case, the tensile zone should be estimated by taking account of the hydration of each thickness as a function of time, or by considering the overall geometry of the element.

To take account of the scale effect in slabs and walls, for which the tensile height is $h_t$ and the spacing between cracks ranges between $h_t$ and 2 $h_t$ (see section 2.2), it is proposed to apply a reduction coefficient taken from Table 6.2 below to the value of the reinforcement section.

In the case of intermediate values of $h_t$, this coefficient may be calculated by interpolating linearly between the associated values in Table 6.2.

In the case of T beams, the results from Table 6.2 may be applied separately to the flange and web regardless of whether or not these have been poured simultaneously. Particular attention should be paid to the reinforcement of the junction between the flange and the web.

| $h_t$ | $f_{ck}$ = 30 MPa | $f_{ck}$ = 40 MPa | $f_{ck}$ = 60 MPa | $f_{ck}$ = 80 MPa |
|---|---|---|---|---|
| 0.015 m | 1.0 | 1.0 | 1.0 | 1.0 |
| 0.30 m | 0.45 | 0.50 | 0.70 | 0.80 |
| ≥ 1.25 m | 0.30 | 0.40 | 0.55 | 0.75 |

**Table 6.2.** *$k_{scale}$ scale effect reduction factors*

NOTE.– The above reduction factors are calculated according to the methodology developed in section 2.2.1.

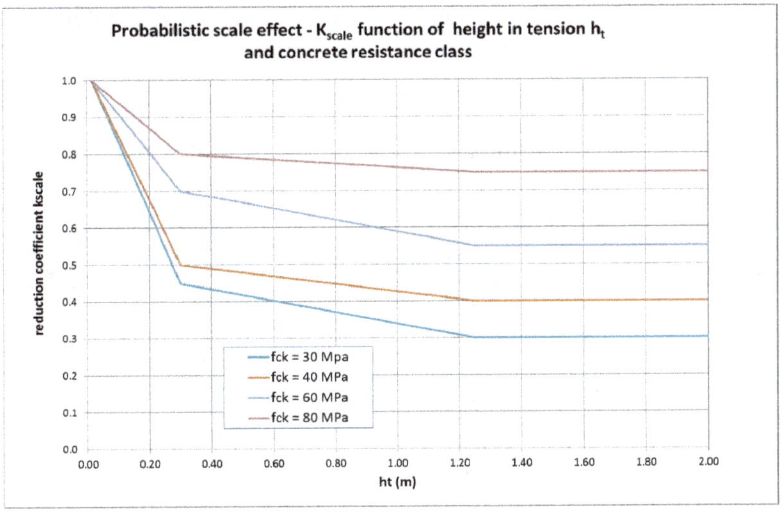


**Figure 6.1.** *Scale effect as a function of the thickness in tension and the compressive strength of the concrete*

### 6.1.2.2. *Long-term reinforcement*

The same principles applied in section 6.1.2.1 may be used to estimate the tensile stresses resulting from long-term effects such as drying or daily temperature cycle.

The following three cases should be considered:

– drying without restraint: in this case, for example a thick reservoir at a suitable distance from the concrete reservoir base, the drying process does not create any mean tensile strength. However, desiccation, which is more important at the surface than at the core, creates tensile strength at the surface;

– drying with restraint: this is typically the case for walls at the base of a building. In general, these walls are subjected to more severe desiccation than the concrete raft, and are therefore considered to be restrained at their base;

– daily variations of temperature of thick elements exposed to the environment.

The principles outlined in section 6.1 for short and long-term aged concrete reinforcement may also be adapted to elements whose thickness is not uniform. In this case, the tensile zone should be estimated by taking account of the drying of each thickness according to time, or by considering the overall geometry of the element.

The reduction factor applied in Table 6.3 is estimated for a given time in the long term.

| *Thickness $h_t$ of the layer in high tension* | *Stress profile* | *Minimum reinforcement for each surface* |
|---|---|---|
| $f_{ctm,\,scale}$ is calculated for $h_t = 0.2\ h/3$ | Drying without external restraint 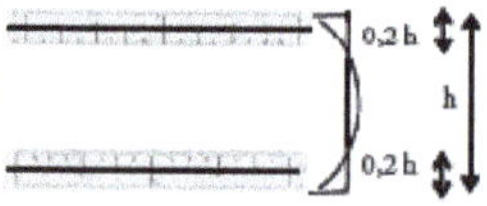  | $A_{smin} = \frac{0.5 \cdot k \cdot 0.2h \cdot 1 f_{ctm,scale}}{f_{yk}}$ |
| $f_{ctm,\,scale}$ is calculated for $h_t = 0.2\ h$ | Drying with end restraint, but excluding edge restraint 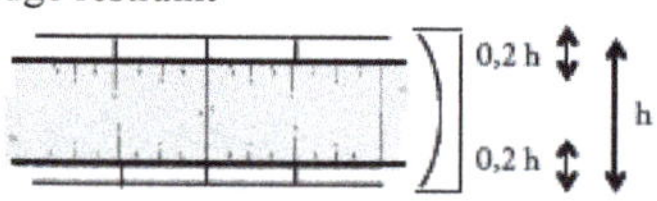  | $A_{smin} = \frac{k \cdot 0.5h \cdot 1 \cdot f_{ctm,scale}}{f_{yk}}$ |
| $f_{ctm,\,scale}$ is calculated for $h_t = 0.3/3 = 0.1$ m | Daily temperature cycle 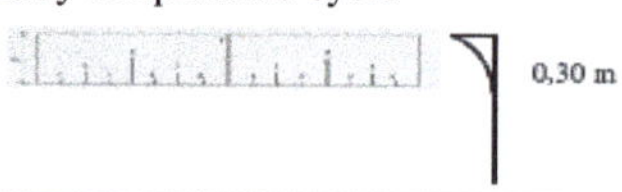  | $A_{smin} = \frac{0.5 \cdot 0.3 \cdot 1 \cdot f_{ctm,scale}}{f_{yk}}$ |

NOTE:

$k$ is the coefficient which allows for the effect of non-uniform self-equilibrating stresses, that lead to reduction of restraint forces

$k = 1$ for webs with $h \leq 300$ mm or flanges with widths less than 300 mm

$k = 0.65$ for webs with $h \geq 800$ mm or flanges with widths greater than 800 mm

$k$ Intermediate values may be interpolated.

**Table 6.3.** *Minimum reinforcement in the long term for concrete thickness h ≥ 400 mm*

## 6.2. Reinforcement of prestressed concrete ties

This section provides a correction to EC2 Part 1-1 equation [7.10] and extends the validation of the equation in the case of stabilized cracking for SLS and ULS calculations.

This correction to [7.10] relates to the insertion of cement grouted multi-strand tendons, in which the tendon diameter is given by: $\emptyset_p = 1.6\sqrt{A_p}$, where $\emptyset_p$ is the diameter of the duct, and is usually chosen to be twice the cross area of the cable.

EN 1992-1-1 section 7.3.4 is applicable to prestressed concrete elements, and takes into account part of the cross-sectional area of the bonded prestressing tendons when estimating the ratio $\rho_{p,eff}$ [7.10]. This expression introduces a percentage $\xi_1^2$ of the cross-sectional area of the bonded tendons:

$$\rho_{p,eff} = \frac{A_s + \xi_1^2 . A_p}{A_{c,eff}}$$

The value of $\xi_1$ is defined in [7.5] as follows: $\xi_1 = \sqrt{\frac{\tau_{bp,k} \cdot \varphi_s}{\tau_{bs,k} \cdot \varphi_p}}$

The ratio of the bond strength between bonded tendons $\tau_{bp,k}$ and ribbed steel of the reinforcing bars in concrete $\tau_{bs,k}$, ($\xi = \frac{\tau_{bp,k \cdot}}{\tau_{bs,k \cdot}}$) is specified in Table 6.2 in section 6.8.2 of EN 1992-1-1.

The theory of cracking in a reinforced concrete or prestressed tie is formulated by [DEB 10] and discussed in section 6.2.2.

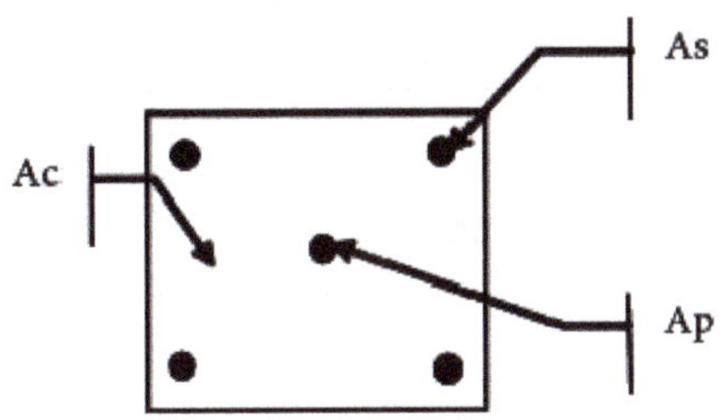


The distribution of stresses between concrete, passive reinforcement and bonded prestressing tendons is illustrated on the figures given in the following for both the crack formation and stabilized cracking phases.

### 6.2.1. *Crack formation in an element in tension*

Figure 6.2 provides the distribution of stresses in steel bars and concrete during crack formation stage (see MC2010 section 7.6.4. Figure 7.6.7 or *fib* bulletin No. 2 Figure 4.3.10).

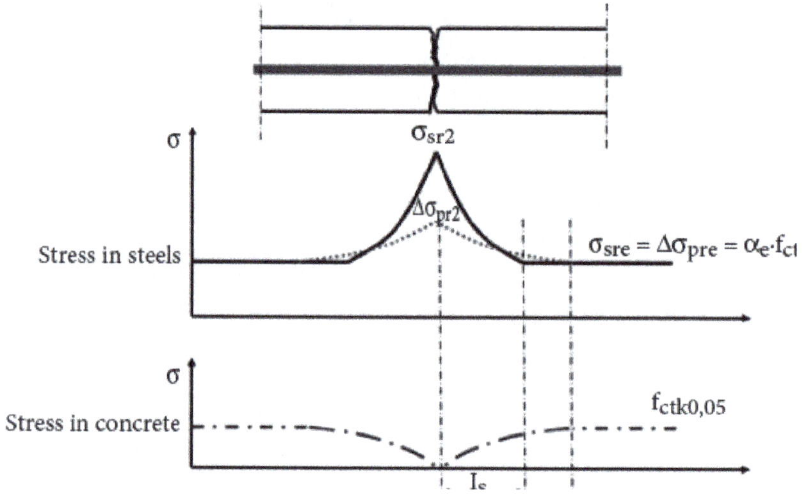


**Figure 6.2.** *Distribution of stresses during crack formation*

| | *Concrete* | *Passive reinforcement* | *Prestressing tendons* |
|---|---|---|---|
| Stress in the cracked section | $\sigma_c = 0$ | $\sigma_{sr2} = \dfrac{N}{(A_s + \sqrt{2}.\xi_1.A_p)}$ | $\Delta\sigma_{pr2} = \dfrac{\sqrt{2}.\xi_1.N}{(A_s + \sqrt{2}.\xi_1.A_p)}$ |
| Stress between cracks | $\sigma_c = \dfrac{N}{\left(A_c + \alpha_e.(A_s + A_p)\right)}$ | $\sigma_{sre} = \dfrac{\alpha_e.N}{\left(A_c + \alpha_e.(A_s + A_p)\right)}$ | $\sigma_{pre} = \dfrac{\alpha_e.N}{\left(A_c + \alpha_e.(A_s + A_p)\right)}$ |

**Table 6.4.** *Concrete and steel stresses during crack formation*

The anchorage length of the passive steel bars is given by:

$$l_s = (\sigma_{sr2} - \sigma_{sre}).\frac{\phi_s}{4.\tau_{bs,k}}$$

The mean strains, $\varepsilon_{sm}$, $\varepsilon_{cm}$ and $\varepsilon_{pm}$, are derived from the stresses, and are estimated over a length of $2.l_s$ (crack formation case).

## 6.2.2. *Stabilized cracking stage in an element in tension*

Figure 6.3 provides the distribution of stresses in steel bars and concrete during stabilized cracking stage (see MC2010 Figure 7.6.-8 or *fib* bulletin no.°2 Figure 4.3.10).

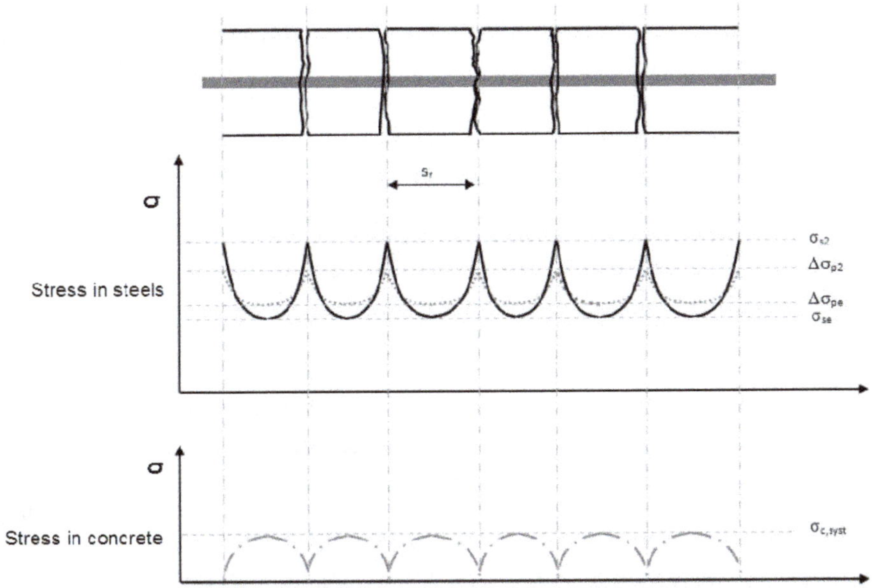


**Figure 6.3.** *Distribution of stresses in steels and concrete*

| | *Concrete* | *Passive reinforcement* | *Prestressing tendons* |
|---|---|---|---|
| Stress in the cracked section | $\sigma_c = 0$ | $\sigma_{s2}$ | $\Delta\sigma_{p2}$ |
| Stress between cracks | $\sigma_{c,syst}$ | $\sigma_{se} = \sigma_{s2} - \dfrac{\sigma_{c,syst}}{(\rho_s + 2.\xi_1^2.\rho_p)}$ | $\Delta\sigma_{pe} = \Delta\sigma_{p2} - 2.\xi_1^2.\dfrac{\sigma_{c,syst}}{(\rho_s + 2.\xi_1^2.\rho_p)}$ |

**Table 6.5.** *Concrete and steel stresses at stabilized cracking stage*

$$\begin{cases} \sigma_{s2} = \dfrac{N}{(A_s+A_p)} + \beta.\sigma_{c,syst}.\left(\dfrac{1}{(\rho_s+2.\xi_1^2.\rho_p)} - \dfrac{1}{(\rho_s+\rho_p)}\right) \\ \Delta\sigma_{p2} = \dfrac{N}{(A_s+A_p)} + \beta.\sigma_{c,syst}.\left(\dfrac{1}{(\rho_s+\rho_p)} - \dfrac{2.\xi_1^2}{(\rho_s+2.\xi_1^2.\rho_p)}\right) \end{cases}$$

where $\sigma_{c,syst}$ is the stress in the concrete corresponding to the formation of the last crack.

$$\sigma_{c,syst} = \frac{N}{\left(A_c+\alpha_e.(A_s+A_p)\right)} = f_{ct,0.95}$$

In the case of stabilized cracking, the stresses in the passive steel reinforcing bars and the prestressing tendons increase as the tensile force N increases. However, the concrete strains and the variations in the strain along the steel reinforcing bars between cracks remain constant.

The spacing between the cracks is given by:

$$S_r = \frac{\phi_s.\sigma_{c,syst}}{2.\tau_{bs,k}.(\rho_s+2.\xi_1^2.\rho_p)}$$

where $S_r = 2.l_S$.

The mean strains, $\varepsilon_{sm}$, $\varepsilon_{cm}$ and $\varepsilon_{pm}$, are derived from the stresses, and are estimated over a length of $S_r$ (stabilized cracking case).

This expression may also be extended to ultimate limit state, assuming a plastic deformation of the passive steel bars. In this case, the stiffness of the section is provided by the stiffness of the prestressing tendons.

### 6.2.3. *Conclusion*

In conclusion, EN 1992-1-1 equation [7.10] should be corrected using $2\zeta_1^2$ in place of $\zeta_1^2$ in the case of a bundle of strands, where the diameter of the equivalent tendon is given by $\emptyset_p = 1{,}6\sqrt{A_p}$. When estimating the stresses in the reinforcement, the contribution of the tendons then becomes $\sqrt{2}\xi_1$ instead of $\xi_1$. Coefficient 2 is the ratio between the cross-sectional area of the duct and the cross-sectional area of the prestressing tendon.

Finally, in the stabilized cracking case the expressions given in section 6.2.2 may be used to justify a larger contribution from the tendons.

## 6.3. Reinforcement of beams

### 6.3.1. *Beams under monotonic mechanical loading*

In the case of simple or T beams, the minimum reinforcement must satisfy the non-fragility condition given by the EC2 section 9.2.1.1 equation or MC2010 section 7.13.5.2 “Beams and T Beams”.

$$A_{s,min} = 0.26 \frac{f_{ctm}}{f_{yk}} b_t d$$

where $b_t$ is the width of the tensile zone.

The above condition also applies to slabs, together with Conditions (2) to (4) as given in EC2 section 9.3.1.1.

In both simple and T beam and slab cases, care should be taken to verify that calculated reinforcement is adequate in accordance with the crack width $w_d$, or requires strengthening.

### 6.3.2. *Beams under imposed deformation and monotonic mechanical loading*

The reinforcement requirements, determined from the mechanical load for massive elements, must be verified for THM effects during early stage and in the long-term.

In performing this verification, the THM effects may be the dominant factor in the combination of mechanical loads, given the THM effects of the corresponding reinforcing bars lie in two different directions or on two different planes.

When it is the case, the resulting minimum reinforcement is considered to be sufficient if it has been calculated taking account of a reduced tensile cross-sectional area and a reduced tensile strength, due to the scale effect as described in section 2.2.

In summary, the proposed method is to verify that the reinforcement is sufficient in the following three successive situations in which cracking must be controlled, under:

– mainly thermal and endogenous effects, at a very early age, and the mechanical effects up to formwork removal, at an age when the concrete has not yet achieved its full properties;

– cumulative thermal and mechanical effects during the various stages from the construction to delivery of the structure to the client before operation up to an age at which the non-fragility can be guaranteed in a concrete that has achieved its full properties;

– thermal, hydric and mechanical effects in long-term service (without cumulating thermal and hydric effects at an early age) during which the concrete develops its relaxation capabilities.

Taking account of the envelope curve of these effects, the same reinforcement can be considered adequate to withstand the various load cases taken into account, which could be considered as independent if the preceding phases have been sufficiently controlled, ensuring that the effects do not result in excessive and irreversible changes to the function of the structure.

## 6.4. Reinforcement of walls

Two cases are to be considered:

– walls without specific cracking requirements;

– walls with specific cracking requirements.

### 6.4.1. *Walls without specific requirements for cracking*

– EC2-1 section 9.6 or MC2010 section 7.13.5.3 apply to reinforced concrete structures;

– for walls fulfilling the criteria specified in Eurocode 2-1 Chapter 12, "plain and lightly reinforced concrete structures", the minimum reinforcement percentage may be zero if there are no other requirements than those relating to cracking (for example, structures in regions subject to earthquakes). The reinforcement percentage may be adjusted to take account of the compression stress. The detailing may also avoid any THM effects.

### 6.4.2. *Walls with specific requirements for cracking*

– For normally thick walls, EC2 or MC2010 detailing applies, avoiding any THM effects;

– for very thick walls, scale effects, according to the approach of Weibull theory and 3D effects, should be taken into account. This results in a reduction to the tensile stress $f_{ctm}$ of the concrete. In practice, it is possible to reduce the tensile stress $f_{ctm}$ by a factor (ranging from 0.65 for 0.80 m thick elements to 0.45 for 2 m thick elements). The values of this reduction factor may be interpolated linearly for intermediate thicknesses. These factors are applicable to walls subject to direct tension, but are not applicable to walls subject to out of plane flexural forces (see MC2010 section 7.13.5.3 “Slabs” for minimum reinforcement specification).

# 7

# Shrinkage, Creep and Other Concrete Properties

## 7.1. Introduction

It should be noted that strains in massive concrete elements are non-uniform across the member cross-section considered.

This chapter considers the shrinkage and creep formulae in MC2010 [CEB 12] and its potential modification in accordance with the conclusions of the test results. These results were performed over a period of at least six months, in accordance with the principles defined in (NF EN 1992-2, 2006) Annex B [NF 06a].

To simplify the approach, modified formulae coefficients are used with a value of 1.0 applied to preliminary studies and values of 0.7 or 1.3 used for detailed studies. Section 7.5 provides a means of determining the modified coefficients in accordance with the experimental results, which inform the coefficients used.

This chapter also considers the French national annex to (NF EN 1992-1-1, 2004 [NF 04]): AN 3 "Application of Annex B", which introduces a three-dimensional approach to stresses and strains. However, recognizing that the creep formulae given in MC2010 differs considerably from those in EN 1992, it is considered that the approach taken in application of the French national annex should be validated.

The drying shrinkage defined in MC2010 is modified to avoid a discontinuity at higher levels of humidity, using a percentage humidity measured at a stabilized stage from Annex B of EN 1992-2.

A dedicated section (section 7.6) considers modifications of the MC2010 parameters to account for concrete element temperatures up to 80°C.

The use of silica fume is considered in EN 1992-2, but is no longer mentioned in MC2010. In the absence of a specific calibration taking into account the use of silica fume in the shrinkage and creep formulae, it remains possible to specify low amplitudes for delayed strains when required by the design. This approach makes reference to the additional requirements given in the NF EN 206/CN standard and detailed further in the French fascicule 65 [SET 14], especially in the case of structures subject to high THM strains and makes use of a performance specification, verified as a minimum at the qualification test stage. Reduced delayed strains may be obtained by the use of silica fume. The shrinkage and creep MC2010 formulae should then be calibrated in accordance with the experimental identification procedure.

One point which has not yet been fully resolved relates to the drying creep in massive concrete elements at ambient temperature. The drying creep formula defined in MC2010 differs from that applied to drying shrinkage, which has much faster kinetics. In contrast to MC2010, the EN 1992-2 annex defines the drying creep as being proportional to the drying shrinkage from the time of loading. Currently, the MC2010 formula is preferred over the EN 1992-2 annex, more convenient when considering that the correction coefficients may be used to obtain a lower kinetic for drying creep.

A further difficulty relates to the extrapolation of measurements on test specimens when applied to massive elements. It is noted that the correction coefficients resulting from tests on specimens are not systematically the same as those obtained from measurements on full-scale structures. However, current measurements on full-scale structures relate to former concrete tests on specimens that are less exhaustive than current tests on specimens. Note, available current tests on specimens relate to full-scale structures which are too recent to be fully used for measurements on these structures [CHA 06].

Feedback from measurements on actual structures similar to those being planned should be taken into account.

## 7.2. Strain

### 7.2.1. *Definition*

The total strain at time *t*, $\varepsilon c(t)$, of a concrete member uniaxially loaded at time $t_0$ with a constant stress $\sigma_c(t_0)$ may be expressed as follows:

$$\varepsilon_c(t) = \varepsilon_{ci}(t_0) + \varepsilon_{cc}(t) + \varepsilon_{cs}(t) + \varepsilon_c T(t)$$

$$\varepsilon_c(t) = \varepsilon_{c\sigma}(t) + \varepsilon_{cn}(t)$$

where:

- $\varepsilon_{ci}(t_0)$ is the initial strain at loading;
- $\varepsilon_{cc}(t)$ is the creep strain at time $t > t_0$;
- $\varepsilon_{cs}(t)$ is the shrinkage strain;
- $\varepsilon_{cT}(t)$ is the thermal strain;
- $\varepsilon_{c\sigma}(t)$ is the stress dependant strain $\varepsilon_{c\sigma}(t) = \varepsilon_{ci}(t_0) + \varepsilon_{cc}(t)$;
- $\varepsilon_{cn}(t)$ is the stress independent strain $\varepsilon_{cn}(t) = \varepsilon_{cs}(t) + \varepsilon_{cT}(t)$.

### 7.2.2. *Range of applicability*

The relations for creep and shrinkage given below predict the time-dependent mean cross-section behavior of a concrete member moist cured at normal temperatures for no longer than 14 days.

Unless special provisions are given, the relations are valid for ordinary structural concrete subjected to a compressive stress $\sigma c \leq 0.4\ f_{cm}(t_0)$ at an age at loading $t_0$ and exposed to mean relative humidity RH in the range of 40 – 100% at mean temperatures from 5°C to 30°C.

The range of concrete strength is:

$$30\text{ MPa} \leq f_{ck} \leq 100\text{ MPa}$$

instead of $15\text{ MPa} \leq f_{cm} \leq 130\text{ MPa}$ in the Model-Code 2010, a range which is too large mainly for thick structures.

The age at loading should be at least one day. Furthermore, the expressions for creep are valid when the mean value of the concrete cylinder strength at the time of loading $f_{cm}(t_0)$ is greater than 0.6 $f_{cm}$ ($f_{cm}(t_0) > 0.6\, f_{cm}$). When concrete is to be loaded at earlier ages, with significant strength development at the beginning of the loading period, specific determination of the creep coefficient should be undertaken. This should be based on an experimental approach.

It is accepted that the relations also apply to concrete in tension, though the relations given in the following paragraphs are directed towards the prediction of creep of concrete subjected to compressive stresses.

### 7.2.3. *Initial strain at loading*

The initial strain $\varepsilon_{ci}(t_0)$ is based on the tangent elasticity modulus, as defined in the following equation (yield curve gradient corresponding to a quick unloading at 0.4 $f_{cm}$):

$$E_{ci} = E_{c0} \cdot \alpha_E \cdot \left(\frac{f_{ck}+8}{10}\right)^{1/3} \qquad [7.1]$$

where:

– $E_{ci}$ is the elasticity modulus (in MPa) for concrete of 28 days;

– $f_{ck}$ is the characteristic strength (in MPa);

– $E_{c0} = 21.5\ 10^3$ MPa.

In the absence of measurements (see section 7.5.1) the parameter $\alpha_E$ as given in the Table 7.1 can be used.

| *Type of aggregate* | $\alpha_E$ |
|---|---|
| Basalt, dense limestone | 1.2 |
| Quartzite | 1.0 |
| Limestone | 0.9 |
| Sandstone | 0.7 |

**Table 7.1.** *Parameter $\alpha_E$ according to aggregate type*

The modulus should be accurately determined when predicting the instantaneous and delayed strains. This is required for special structures (particularly for preventing thermal cracking), where an experimental identification procedure should be considered.

For example, the revised French regulations (see fascicule 65, section 8.1.1.5 "Additional data") should be used.

NOTE 7.1.– Concrete modulus, shrinkage and creep tests are included in qualification tests. Other modulus tests (namely information test, fascicule 65, section 8.3.2.4 "information tests") are required during the concrete production phase to check the value of concrete modulus.

For a mean temperature of 20°C and curing in accordance with ISO 1920-3 [ISO 04] and EN 13670 [NF 13], the relevant compressive strength of concrete at various ages $f_{cm}(t)$ may be estimated by the following equations:

$$f_{cm}(t) = \beta_{cc}(t) \cdot f_{cm} \quad [7.2]$$

$$\beta_{cc}(t) = \exp\left\{s \cdot \left[1 - \left(\frac{28}{t}\right)^{0,5}\right]\right\}. \quad [7.3]$$

where "s" is given by the following Table 7.2:

| $f_{cm}$ *(MPa)* | *Strength class of cement* | *s* |
|---|---|---|
| ≤ 60 | 32.5 N | 0.38 |
| | 32.5 R; 42.5 N | 0.25 |
| | 42.5 R; 52.5N; 52.5R | 0.20 |
| >60 | All classes | 0.20 |

**Table 7.2.** *s value according to cement strength class*

The elasticity modulus of concrete at an age t ≠ 28 days may be estimated from:

$$E_{ci}(t) = [\beta_{cc}(t)]^{0.5} \cdot E_{ci} \quad [7.4]$$

For the prediction of the creep function, the initial strain $\varepsilon_{ci}(t_0)$ is based on the tangent elasticity modulus defined as follows:

$$\varepsilon_{ci}(t_0) = \frac{\sigma_c(t_0)}{E_{ci}(t_0)} \quad [7.5]$$

A more accurate estimate can be used by taking into account each component of the concrete stresses and the Poisson's ratio.

## 7.3. Shrinkage

The total shrinkage or swelling strains $\varepsilon_{cs}(t,t_s)$ may be calculated from the following equation:

$$\varepsilon_{cs}(t,t_s) = \beta_{ca1} \cdot \varepsilon_{cas}(t) + \beta_{cd1} \cdot \varepsilon_{cds}(t,t_s) \quad [7.6]$$

where shrinkage is subdivided into:

– the autogenous shrinkage $\varepsilon_{cas}(t)$:

$$\varepsilon_{cas}(t) = \varepsilon_{cas0}(f_{cm}) \cdot \beta_{as}(t) \quad [7.7]$$

– and the drying shrinkage $\varepsilon_{cds}(t,t_s)$:

$$\varepsilon_{cds}(t,t_s) = \varepsilon_{cds0}(f_{cm}) \cdot \beta_{RH}(RH) \cdot \beta_{ds}(t - t_s) \quad [7.8]$$

where

– $t$ is the concrete age (in days);

– $t_s$ is the concrete age at the beginning of drying (in days);

– $(t\text{-}t_s)$ is the duration of drying (in days).

Where actual shrinkage measurements are not available (see section 7.5.2), the parameters $\beta_{ca1}$ and $\beta_{cd1}$ are taken as equal to 1.0 for preliminary studies and 0.7 or 1.3, depending upon either the favorable or unfavorable effects of the imposed deformation, for detailed studies. It is recommended that the parameter value used is identified experimentally prior to construction.

### 7.3.1. *Autogenous shrinkage*

Shrinkage component $\varepsilon_{cas}(t)$ may be estimated by means of the notional autogenous shrinkage coefficient $\varepsilon_{cas0}(f_{cm})$ and the time function $\beta_{as}(t)$:

$$\varepsilon_{cas0}(f_{cm}) = -\alpha_{as}\left(\frac{f_{cm}/_{10}}{6+f_{cm}/_{10}}\right)^{2,5} \cdot 10^{-6} \qquad [7.9]$$

$$\beta_{as}(t) = 1 - exp\left(-0,2 \cdot \beta_{ca2}\sqrt{t}\right)$$

where:

– $f_{cm} = f_{ck} + 8$ MPa (MC2010 Eq. [5.1.1]), and $fcm$ is the mean compressive strength in [MPa] at a concrete age of 28 days;

– $\alpha_{as}$ is a coefficient, dependent upon the type of cement (see Table 7.3):

| Strength class of cement | $\alpha_{as}$ |
|---|---|
| 32,5 N | 800 |
| 32,5 R; 42,5 N | 700 |
| 42,5 R; 52,5N; 52,5R | 600 |

**Table 7.3.** *Coefficient α according to cement strength class*

Where actual autogenous shrinkage measurements are not available (see section 7.5.2) the parameter $\beta_{ca2}$ is taken as equal to 1.0 for preliminary studies and to 0.7 or 1.3, depending upon either the favorable or unfavorable effects of the imposed deformation, for detailed studies. It is recommended that the parameter value used is identified via experimental means prior to construction.

NOTE 7.2.– the time period, (in days), when half the autogenous shrinkage is reached is equal to $12/\beta_{ca2}{}^2$ (in days).

### 7.3.2. *Drying shrinkage*

The drying shrinkage $\beta_{cds}(t,t_s)$ is calculated using the notional drying shrinkage coefficient $\varepsilon_{cds0}(f_{cm})$, the coefficient $\beta_{RH}(RH)$, taking into account the effect of the ambient relative humidity, and the function $\beta_{ds}(t-t_s)$ describing the time-development:

$$\varepsilon_{cds0}(f_{cm}) = [(220 + 110\alpha_{ds1}) \cdot exp(-\alpha_{ds2} \cdot f_{cm})] \cdot 10^{-6} \quad [7.10]$$

$$\beta_{RH} = -1.55\left[1 - \left(\frac{RH}{RH_{eq}}\right)^3\right] \text{ for } 40 \leq RH \leq RH_{eq} \quad [7.11]$$

$$\beta_{RH} = -1.55\left[1 - \left(\frac{RH}{RH_{eq}}\right)^2\right] \text{ for } RH \geq RH_{eq}$$

where with $RH_{eq} = 72 \exp[-0.046 (f_{cm} - 8)] + 75$

$$\beta_{ds}(t - t_s) = \left(\frac{(t-t_s)}{\beta_{cd2} \cdot 0.035 \cdot h^2 + (t-t_s)}\right)^{0,5} \quad [7.12]$$

Where actual measurements are not available (see section 7.5.2) the parameter $\beta_{cd2}$ is taken as equal to 1.0 for preliminary studies and to 0.7 or 1.3, depending upon either the favorable or unfavorable effect of the imposed deformation, for detailed studies. It is recommended that the parameter value used is identified via experimental means prior to construction.

where

– $\alpha ds1$, $\alpha ds2$ are coefficients, dependent on the type of cement (see Table 7.4):

| Strength class of cement | $\alpha_{ds1}$ | $\alpha_{ds2}$ |
|---|---|---|
| 32.5 N | 3 | 0.013 |
| 32.5 R; 42.5 N | 4 | 0.012 |
| 42.5 R; 52.5N; 52.5R | 6 | 0.012 |

**Table 7.4.** *Coefficients αds1 and αds2 according to the cement strength classes*

– $f_{cm} = f_{ck} + 8$ MPa (MC2010 Eq. 5.1-1), where $f_{cm}$ is the mean compressive strength in MPa at a concrete age of 28 days;

– *RH* is the relative humidity, in [%], of the ambient atmosphere in [%];

– $h = 2Ac/u$ is the notional size of member in mm, with $A_c$ as the cross-section in mm$^2$ and $u$ as the perimeter of the member in contact with the atmosphere in mm;

– $t$ is the concrete age in days;

– $t_s$ is the concrete age at the beginning of drying in days;

– $(t-t_s)$ is the duration of drying in days.

For special structures, consideration should be given to the level of approximation used; this approximation results from the section calculation and is based on the notional member size. This is especially applicable in the case of structure zones, where components of various thicknesses are connected together. In this case, it is more appropriate to firstly perform a drying calculation, followed by a mechanical calculation with strain values dependent upon the saturation or relative humidity parameters [CHA 06]. This approach typically applies to box-girder bridges, where the thickness of top slab and box-girder flanges varies significantly, which in turn generates strain redistributions due to the difference of shrinkage and creep kinetics.

The proposed $\beta_{RH}(RH)$ value is modified taking into account the MC2010 approach, as follows:

$$\beta_{RH} = -1.55\left[1-\left(\frac{RH}{100}\right)^3\right] \text{ for } 40 \leq RH \leq 99\%\beta_{s1} \qquad [7.13]$$

$$\beta_{RH} = 0.25 \text{ for } RH \geq 99\%\beta_{s1}$$

$$\beta_{s1} = (35/fcm)^{0.1}$$

This MC2010 modification avoids the discontinuity for $99\%\beta_{s1}$ and over shrinkage from RH = 80%, as verified from existing cooling tower measurements. Figure 7.1 illustrates the following comparison between equations:

– $\varepsilon_{cdso}(f_{cm}) \times \beta_{RH}(RH)$ (MC2010)

– $K(f_{ck})$ {72 exp [0,046 $(f_{cm}$ - 8)] + 75 – RH} (EN1992-2)

for four compressive strength values between $f_{ck}$ = 30 MPa and $f_{ck}$ = 70 MPa:

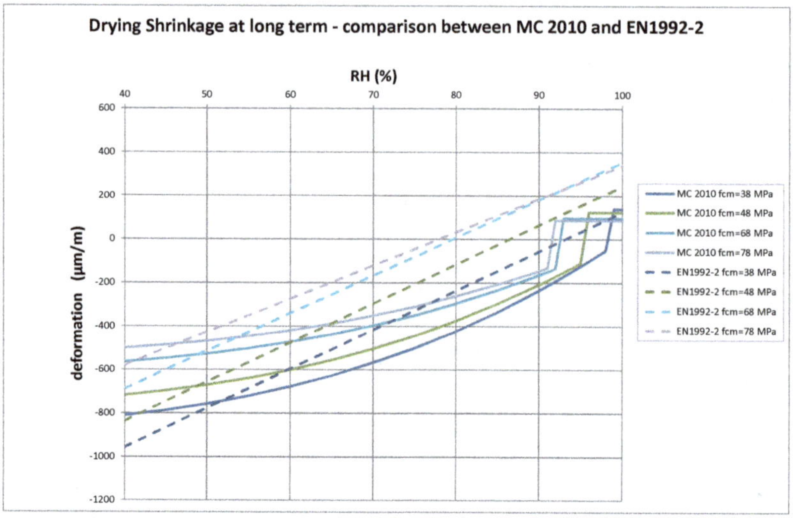


**Figure 7.1.** *Long-term drying shrinkage – comparison between MC2010 and EC2 (EN 1992-2)*

Figure 7.2 gives the application of the above-modified equations. The distribution of the graph results are closely correlated to MC2010 values between RH = 40% and RH = 70%, and EN 1992-2 values between RH = 80% and RH = 100%. Associated discontinuities have been removed.

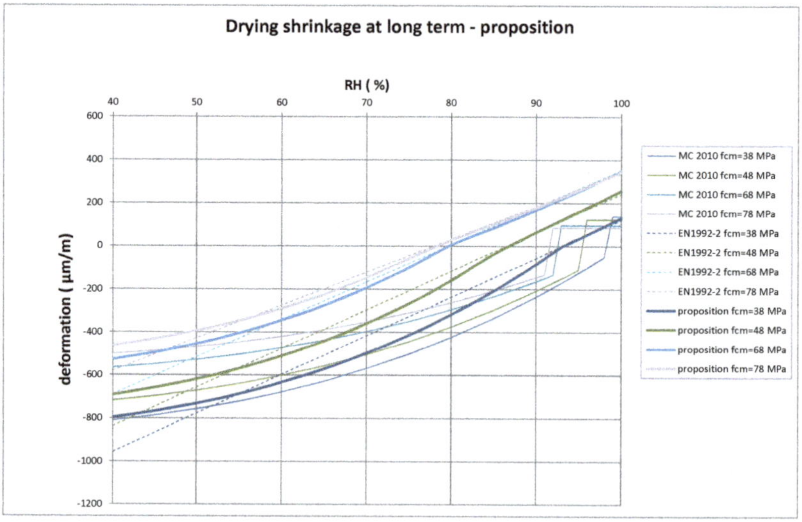


**Figure 7.2.** *Long-term drying shrinkage – proposal*

NOTE 7.3.– the period that is the time when half of the drying shrinkage is reached is equal to $0.035h^2/3$ days, that is 11667 days (approximately 32 years) when h = 1 m, and 1867 days when h = 0.40 m, in the case when $\beta_{cd2} = 1.0$.

## 7.4. Creep

### 7.4.1. *Assumptions and related basic equation*

Within the range of service stresses

$$|\sigma_c| \leq 0{,}4 \cdot f_{cm}(t_0)$$

creep is assumed to be linearly related to stress.

The basic creep is considered an anisotropic deformation, and the basic creep deformation as proportional to the instantaneous deformations $\varepsilon_{ci,x}$ in the direction under consideration.

The drying creep deformation is isotropic and proportional to $\sigma_{cm}/E_{ci} = (\sigma_1+\sigma_2+\sigma_3)/E_{ci}$, with $\sigma_1$, $\sigma_2$ and $\sigma_3$ representing the principal stresses.

For a constant stress applied at time $t_0$ this leads to:

$$\varepsilon_{cc}(t,t_0) = \varepsilon_{ci} \cdot \boldsymbol{\beta_{bc1}} \cdot \beta_{bc}(f_{cm}) \cdot \beta_{bc}(t,t_{0,adj}) + \frac{\sigma_{cm}}{E_{ci}} \cdot \beta_{dc1} \cdot \beta_{dc}(f_{cm}) \cdot \beta(RH) \cdot \beta_{dc}(t_{0,adj}) \cdot \beta_{dc}(t,t_0) \quad [7.14]$$

Where actual measurements are not available (see section 7.5.3) the parameters $\beta_{bc1}$ and $\beta_{dc1}$ are taken as equal to 1.0 for preliminary studies and 0.7 or 1.3, depending upon either the favorable or unfavorable effects of the imposed deformation, for detailed studies. It is recommended that the parameter values used are identified experimentally prior to construction.

where:

– $E_{ci}$ is the elasticity modulus at the age of 28 days according to equations in section 7.2.2;

– $\beta_{bc}(t,t_0)$ is the coefficient to describe the development of basic creep according to time after loading;

– $\beta_{dc}(t,t_o)$ is the coefficient to describe the development of drying creep according to time after loading;

– $t$ is the age of concrete (in days) at the considered moment;

– $t_{o,adj}$ is the age of concrete at loading, (in days), adjusted in accordance with equation [7.15].

The effects from the type of cement on the concrete creep coefficient may be taken into account by modifying the age at loading $t_{0,T}$ $t_{0,adj}$ $t$ in accordance with the following equation:

$$t_{o,adj} = \beta_{bc2} \cdot t_{0.T} \cdot \left[\frac{9}{1+t_{0,T}^{1.2}} + 1\right]^{\alpha} \geq 0.5\ days \qquad [7.15]$$

where:

– $\beta_{bc2}$ is an adjustment coefficient derived from experimental tests;

– $t_{0,T}$ is the concrete loading age in days, adjusted according to the equation given in section 7.6.2, and $\alpha$ in accordance with the values given in Table 7.5.

| *Strength class of cement* | $\alpha$ |
|---|---|
| 32.5 N | -1 |
| 32.5 R; 42.5 N | 0 |
| 42.5 R; 52.5N;52.5 R | +1 |

**Table 7.5.** *Coefficient $\alpha$ according to cement strength class*

NOTE 7.4.– where actual creep measurements are not available (see section 7.5.3) the parameter $\beta_{bc2}$ is taken as equal to 1.0 for preliminary studies and to 0.7 or 1.3, depending upon either the favorable or unfavorable effects of the imposed deformation, for detailed studies. It is recommended that the parameter value used is identified via experimental means prior to construction.

### 7.4.2. *Basic creep*

The basic creep coefficient may be estimated using the following equation:

$$\varphi_{bc}(t,t_0) = \beta_{bc}(f_{cm}) \cdot \beta_{bc}(t,t_0) \qquad [7.16]$$

with

$$\beta_{bc}(f_{cm}) = \frac{1{,}8}{(f_{cm})^{0.7}}$$

where $f_{cm}$ = $f_{ck}$+ 8 MPa (MC2010 Eq. [5.1.1]), and $f_{cm}$ is the mean compressive strength in MPa at a concrete age of 28 days.

The development of basic creep with time is described by MC2010 Eq. [5.1.1]:

$$\beta_{bc}(t,t_0) = ln\left[\left(\frac{30}{t_{0,adj}} + 0.035\right)^2 \cdot (t-t_0) + 1\right] \qquad [7.17]$$

### 7.4.3. *Drying creep*

The drying creep coefficient may be estimated using the following equation:

$$\varphi_{dc}(t,t_0) = \beta(RH) \cdot \beta_{dc}(f_{cm}) \cdot \beta_{dc}(t_{0,adj}) \cdot \beta_{dc}(t,t_0) \qquad [7.18]$$

with

$$\beta(RH) = \frac{1-\frac{RH}{100}}{\sqrt[3]{0.1 \cdot \frac{h}{100}}}$$

$$\beta_{dc}(f_{cm}) = \frac{412}{(f_{cm})^{1,4}}$$

$$\beta_{dc}(t_0) = \frac{1}{0.1+(t_{0,adj})^{0,2}}$$

where:

$-f_{cm}$ = $f_{ck}$+ 8 MPa (MC2010 Eq. [5.1.1]), where $f_{cm}$ is the mean compressive strength in MPa at a concrete age of 28 days;

– *RH* is the ambient environment relative humidity in [%];

– $h = 2Ac/u$ = the notional member size in mm, where *Ac* is the cross-section in mm² and *u* is the perimeter of the member in mm in contact with the atmosphere.

The development of drying creep with time is given by:

$$\beta_{dc}(t, t_0) = \left[\frac{t - t_0}{\beta_{dc2} \cdot \beta_h + t - t_0}\right]^{\gamma(t_0)} \quad [7.19]$$

Where actual measurements are not available (see section 5.5.5) the parameters $\beta_{dc2}$ is taken as equal to 1.0 for preliminary studies and 0.7 or 1.3 depending upon either the favorable or unfavorable effects of the imposed deformation, for detailed studies. It is recommended that the parameter value used is identified experimentally prior to construction.

with:

$$\beta_h = 1.5 \cdot h + 250\alpha_{fcm} \leq 1500\alpha_{fcm} \quad [7.20]$$

$$\gamma(t_{0,adj}) = \frac{1}{2.3 + \frac{3.5}{\sqrt{t_{0,adj}}}}$$

$\gamma(t_{0,adj})$ ranges from 0.3 to 0.4 for loading times of 14 days or one year respectively.

Where

$$\alpha_{f_{cm}} = \left[\frac{35}{f_{cm}}\right]^{0.5} \leq 1 \quad [7.21]$$

For the kinetic formula of drying creep evolution, the exponent used is different from the exponent for drying shrinkage:

$\gamma(t_{0,adj})$ = 0.3 to 0.4 for creep and 0.5 for shrinkage.

Similarly, the coefficient characterizing the drying shrinkage kinetic refers to $h^2$:

$$\left(\frac{(t - t_s)}{\beta_{cd2} \cdot 0.035 \cdot h^2 + (t - t_s)}\right)^{0,5} \quad [7.22]$$

and the coefficient characterizing the drying creep kinetic is approximately proportional to h:

$$\beta_{cd2} \cdot min\{1.5h + 250\alpha_{fcm}\,; 1500\alpha_{fcm}\} \qquad [7.23]$$

Note, consideration should be given to the difference between the drying shrinkage and drying creep kinetics. This leads to separate periods being considered for drying shrinkage and drying creep.

EXAMPLE.– For a 40 MPa strength class concrete with a thickness h = 1000 mm, the drying shrinkage period is 11000 days and the drying creep period is 190 days.

NOTE 7.5.– The period, the time when half of the drying creep is reached, is equal to:

$$\beta_{dc2} \cdot \beta_h / \left[2^{1/\gamma(t_{0,adj})} - 1\right] \qquad [7.24]$$

At this stage of the design the MC2010 formula may be used, given that through the use of correcting coefficients, it is possible to obtain a slower kinetic for the drying creep.

As for the drying shrinkage calculation (see section 7.3.2), in the case of special structures, consideration should be given to the level of approximation used, which results from the calculation of the section based on the notional size of member. This is especially applicable in the case of structure zones where components of various thicknesses are connected together.

## 7.5. Experimental identification procedures

When evaluating delayed strains to a greater level of accuracy, it may be necessary to modify the model parameters for creep and shrinkage by using experimental measurements. The following procedure may be used; this modifies the above formulae (sections 7.3 and 7.4) and is based upon the use of experimental results, which inform the coefficients used.

### 7.5.1. *Initial strain at loading time*

Experimental data may be obtained from the Young's modulus test measured according to NF EN 12390-13.

The parameter $\alpha_E$ should be chosen in order to minimize the sum of the square of the differences between the model estimate and the experimental results.

Experimental data may be obtained from appropriate shrinkage and creep tests, both in autogenous and drying conditions. The measurements should be obtained under controlled conditions and recorded for at least six months' duration.

### 7.5.2. *Shrinkage*

The two parameters $\beta_{ca1}$ and $\beta_{cd1}$,, which give the asymptote of each part of shrinkage, and the two parameters $\beta_{ca2}$ and $\beta_{cd2}$, which give the kinetic of each part of shrinkage, should be chosen in order to minimize the sum of the square of the difference between the model estimate and the experimental results.

### 7.5.3. *Basic creep*

The two parameters $\beta_{bc1}$ and $\beta_{bc2}$ should be determined by minimizing the sum of the square of the difference between the model estimate and experimental results.

### 7.5.4. *Drying creep*

The two parameters $\beta_{dc1}$ and $\beta_{dc2}$ should be determined by minimising the sum of the square of the difference between the model estimate and experimental results.

### 7.5.5. *Estimation of long-term delayed strain*

Creep and shrinkage formulae and experimental output are based on data collected over limited periods of time. Extrapolating their use to longer-term applications (i.e. 100 years) results in the introduction of additional errors, associated with the mathematical expressions used in the extrapolation.

The following formulae provide an acceptable estimated average for delayed strains when extrapolated to the long term. However, when a further margin of safety is introduced by the overestimation of delayed strains and when it is applicable to the design, the creep and shrinkage predicted on the basis of the formulae or experimental output should be multiplied by a safety factor.

In order to take into account uncertainties in the actual long-term delayed concrete strains (uncertainties in the validity of extrapolating the use of mathematical formulae for applying creep and shrinkage measurements to a relatively short time period), the following safety factor $\gamma_{lt}$ may be applied;

where

$t \leq 1.0$ year $\gamma_{lt} = 1$

$t \geq 1.0$ year $\gamma_{lt} = 1 + 0.1\, log\left(\frac{t}{t_{ref}}\right)$ with $t_{ref} = 1.0$ year. [7.25]

Where experimental outputs are based on existing structure long-term data, the safety factor $\gamma_{lt}$ can be equal to 1.0.

## 7.6. Temperature effects on concrete properties

The following information is taken from MC2010. Note, some concrete characteristics, such as strength and elasticity modulus, are subject to significant change from 20°C when EN 1992-1-2 and EN 19992-3 only consider changes >50°C and even potentially beyond 100°C.

For example, EN 1992-1-2 does not consider any change of concrete compressive or tensile strength up to 100°C. Only $\varepsilon_{c1}$ and $\varepsilon_{cu1}$ changes are considered beyond 20°C.

EN 1992-3 Annex K states that the concrete modulus is not intended to vary up to temperatures of 50°C and that the creep coefficient increases from 1.0 to 1.35 over a range of 20°C to 50°C for hot element loading.

The validity of characteristic changes beyond 20°C should be verified by appropriate measurements.

### 7.6.1. *Temperature effects on instantaneous concrete characteristics*

The information given in the clauses in section 7.5 is valid for a mean concrete temperature, taking into account seasonal variations, between approximately –20°C and +40°C. In the following clause, the effect of substantial deviations from a mean concrete temperature of 20°C in the range of approximately 0°C to +80°C is considered.

### 7.6.2. *Maturity*

The effect of elevated or reduced temperatures on the maturity of concrete may be taken into account by adjusting the concrete age, according to the following equation:

$$t_T = \sum_{i=1}^{n} \Delta t_i \, exp\left[13.65 - \frac{4000}{273+T(\Delta t_i)}\right] \quad [7.26]$$

where

– $t_T$ is the temperature adjusted concrete age (or effective concrete age) which replaces $t$ in the corresponding equations (in days);

– $\Delta t_i$ is the number of days where a temperature $T$ prevails;

– $T(\Delta t_i)$ is the temperature in °C during the time period $\Delta t_i$.

### 7.6.3. *Thermal expansion*

The coefficient of thermal expansion is dependent on the type of aggregates used, the moisture content of the concrete and may vary between approximately $6 \cdot 10^{-6}$ K $^{-1}$ and $15 \cdot 10^{-6}$ K$^{-1}$.

For design, a coefficient of thermal expansion value of $10 \cdot 10^{-6}$ K $^{-1}$ may be taken for normal strength and high strength concrete.

The thermal expansion of concrete may be calculated using the following equation:

$$\varepsilon_{cT} = \alpha_T \cdot \Delta T \quad [7.27]$$

where:

– $\varepsilon_{cT}$is the thermal strain;

– $\Delta T$*is* the change of temperature in (K);

– $\alpha_T$is the coefficient of thermal expansion in ($K^{-1}$).

### 7.6.4. *Compressive strength*

The effect of temperature, in the range of $0°C \leq T \leq 80°C$, on the compressive strength, $f_{cm}(T)$, of normal strength and high strength normal weight aggregate (see MC2010 and EN 206) concrete may be calculated using the following equation:

$$f_{cm}(T) = f_{cm}(1.06 - 0.003 \cdot T) \qquad [7.28]$$

where:

– $f_{cm}(T)$ is the compressive strength in [MPa] at the temperature Tin °C;

– $f_{cm}$is thecompressive strength in [MPa] at the temperature 20°C;

– Tis the temperature in (°C).

The concrete compressive strength reduction is 6 % at 40°C and 12 % at 60°C.

### 7.6.5. *Tensile strength*

The effect of temperature in the range of $0°C \leq T \leq 80°C$ on the compressive strength of normal strength and high strength normal weight aggregate concrete, $f_{cm}(T)$, may be calculated from the following equation:

$$f_{cm}(T) = f_{cm} . (1.16 - 0.008 . T) \qquad [7.29]$$

where:

– $f_{cm}(T)$ is the compressive strength in MPa at the temperature T in °C;

– $f_{cm}$ is the compressive strength in MPa at the temperature 20°C;

– $T$ is the temperature in °C.

The concrete tensile strength reduction is 16 % at 40°C and 32 % at 60°C.

### 7.6.6. *Fracture energy*

Fracture energy is significantly affected by temperature and moisture content at the time of testing. The effect of temperature on the fracture energy $G_F$ (T) of normal strength normal weight concrete may be estimated using the following equations:

dry concrete: $G_F(T) = G_F \, . \, (1.06 - 0.003\ T)$

mass concrete: $G_F(T) = G_F \, . \, (1.12 - 0.006\ T)$

where:

– $G_F(T)$ is the fracture energy in N/m at a temperature $T$ in °C;

– $G_F$ is the fracture energy in N/m at a temperature of 20°C.

Where no experimental data are available, fracture energy in N/m at 20°C for ordinary normal weight concrete may be estimated using the following equation:

$$G_F = 73\ (f_{cm})^{0,18} \qquad [7.30]$$

The concrete fracture energy reduction is 12 % at 40°C and 24 % at 60°C for mass concrete.

### 7.6.7. *Elasticity modulus*

The effect of elevated or reduced temperatures at the time of testing on the elasticity modulus of normal strength and high strength normal weight concrete, at an age of 28 days, may be estimated using the following equation:

$$E_{ci}(T) = E_{c0} \cdot \alpha_E \cdot \left(\frac{f_{ck}+8}{10}\right)^{1/3} \cdot (1.06 - 0.003 \cdot T) \qquad [7.31]$$

where $E_{ci}(T)$ is the elasticity modulus in MPa at temperature $T$ in C°.

The concrete modulus reduction is 6% at 40°C and 12% at 60°C.

### 7.6.8. *Temperature effects on the delayed deformations*

When considering temperature effects on delayed deformations, three aspects should be considered:

– an acceleration of the drying kinetic related to associated strains. This results mainly from the increase of the diffusion coefficient, according to temperature *T* (outside humidity equilibrium is normally different and may be taken into account independently);

– an increase of the range of basic creep;

– potential effects on the maturity of autogenous shrinkage.

#### 7.6.8.1. *Temperature effect prior to loading*

The temperature effects prior to loading may be taken into account by adjusting the concrete age in accordance with the effective concrete age $t_T$ (see section 7.6.2).

#### 7.6.8.2. *Temperature effect during loading*

The following equations consider the temperature effect on the drying creep development time, where the constant temperature differs from 20°C for a normal weight concrete under load. The effect of drying creep development time is taken into account by using the factor $\beta_T$ in the following equation:

$$\beta_{hT} = \beta_h \, . \, \beta_T \qquad [7.32]$$

with $\beta_T = \exp\,[1500 / (273 + T) - 5.12]$

where:

– $\beta_{hT}$ is a temperature dependent coefficient replacing $\beta_h$;

– $\beta_h$ is the coefficient without temperature effect $\beta_h = 1.5 \cdot h + 250\alpha_{fcm} \leq 1500\alpha_{fcm}$;

– *T* is the temperature in °C.

The acceleration of drying kinetic is reflected by a reduction of coefficient $\beta_h$, equal to 28% at 40°C and 46% at 60°C.

The effect of temperature on the creep coefficient is taken into account using the following equations:

$$\varphi_{bc,T} = \varphi_{bc} \cdot \varphi_T$$

$$\varphi_{dc,T} = \varphi_{dc} \cdot \varphi_T^{1.2} \quad [7.33]$$

with

$$\varphi_T = \exp\,[\,0.015\,(\,T - 20)\,]$$

where:

- $\varphi_{bc,T}$ is a temperature dependent coefficient which replaces $\varphi_{bc}$;
- $\varphi_{dc,T}$ is a temperature dependent coefficient which replaces $\varphi_{dc}$;
- $\varphi_{bc}$ is the basic creep coefficient without temperature effect:

$$\varphi_{bc}(t, t_0) = \beta_{bc}(f_{cm}) \cdot \beta_{bc}(t, t_0);$$

- $\varphi_{dc}$ is the basic creep coefficient without temperature effect.

$$\varphi_{dc}(t, t_0) = \beta(RH) \cdot \beta_{dc}(f_{cm}) \cdot \beta_{dc}(t_{0,adj}) \cdot \beta_{dc}(t, t_0) \quad [7.34]$$

- $T$ is the temperature in °C.

The increase of basic creep range is from 35% at 40°C to 82% at 60°C.

The increase of drying creep range is from 43% at 40°C to 105% at 60°C.

### 7.6.8.3. *Temperature effect post loading*

For a temperature increase while the structural member is under load, creep may be estimated using the following equation:

$$\varphi(t, t_0, T) = \varphi(t, t_0) + \Delta\varphi_{T,trans} \quad [7.35]$$

with

$$\Delta\varphi_{T,trans} = 0.0004(T - 20)^2$$

where:

– $\varphi(t,t_0)$ is the creep coefficient taking into account the temperature adjustment:

$$\varphi(t,t_0) = \varphi_{bc}(t,t_0) + \varphi_{dc}(t,t_0); \qquad [7.36]$$

– $\Delta\ \varphi_{T,trans}$ is the transient thermal creep coefficient which occurs at the time of the temperature increase;

– $T$ is the temperature in °C.

The creep coefficient increase is from 0.16 at 40°C to 0.64 at 60°C. The increase of this coefficient and should be separated between basic creep and drying creep. At the present stage of the project, no clear approach to separation has been identified.

### 7.6.9. *Autogenous shrinkage*

Temperatures between 0°C and 80°C mainly influence the time development of autogenous shrinkage. Therefore, the autogenous shrinkage at a concrete age $t$ is calculated using the effective concrete age $t_T$:

$$\varepsilon_{cas}(t) = \varepsilon_{caso}(f_{cm}) \cdot \beta_{as}(\boldsymbol{t_T}) \qquad [7.37]$$

### 7.6.10. *Drying shrinkage*

The following equations consider the effect of a constant temperature, which differs from 20°C while the concrete is drying:

The temperature effect on the time development of drying shrinkage is taken into account replacing the product 0.035 $h^2$ by 0.035. $h^2$. exp [-0.06(T-20)]:

$$\beta_{ds}(t - t_s) = \left(\frac{(t-t_s)}{\beta_{cd2} \cdot 0.035 \cdot h^2 \cdot exp[-0.06(T-20)] + (t-t_s)}\right)^{0.5} \qquad [7.38]$$

The acceleration of the drying shrinkage kinetic results in a reduction of the coefficient $0.035h^2$ of 70% at 40°C and 91% at 60°C.

Further to section 7.4.2, it should be noted that the shrinkage and creep drying accelerations are different. However, it is proposed to keep the same relationship for both.

The effect of temperature on the notional shrinkage coefficient is taken into account as follows, using the following parameters $\beta_{sT}$ and $\beta_{s1T}$:

$$\varepsilon_{cds}(t,t_s) = \varepsilon_{cds0}(f_{cm}) \cdot \beta_{RH}(RH) \cdot \beta_{sT} \cdot \beta_{ds}(t-t_s) \qquad [7.39]$$

$$\beta_{RH} = -1.55\left[1-\left(\frac{RH}{RH_{eq}}\right)^3\right] \text{ for } 40 \leq \text{RH} \leq \text{RH}_{\text{eq,T}}$$

$$\beta_{RH} = -1.55\left[1-\left(\frac{RH}{RH_{eq}}\right)^2\right] \text{ for RH} \geq \text{RH}_{\text{eq,T}}$$

$$RH_{eq,T} = 72 \cdot \exp[0.046(f_{cm} - 8)] + 75 + \beta_{s1,T}$$

where:

$$\beta_{s,T} = 1 + \left(\frac{4}{103-RH}\right)\left(\frac{T-20}{40}\right) \qquad [7.40]$$

$$\beta_{s1,T} = \left(\frac{T-20}{25}\right)^3.$$

The increase in drying shrinkage range is from 4% at 40°C, with an ambient relative humidity of 50%, to 8% at 60°C with a relative humidity of 50%.

# 8

# Cracking of Beams and Walls Subject to Restrained Deformations at SLS

Two types of cracking, due to early age concrete shrinkage, may occur in massive elements:

– cracking on a structure surface resulting from the temperature difference between the concrete core and the surface;

– cracking due to the bulk cooling of the concrete element, including elements with:

- an edge restrained on one side. This is typically the case for a raft cast on soil, or a fresh concrete lift for a wall placed on top a previous lift, or on a raft;

- an end restrained. This is typically the case for a concrete slab constrained between two rigid walls.

Cracking on the concrete surface can normally be avoided by adopting a number of precautions, including the thermal insulation characteristics of formworks and by an adequate curing period. This is required in order to limit the temperature difference between the concrete core and surface.

When surface cracking cannot be avoided, reinforcement of the concrete surface allows limitation of crack width.

## 8.1. Evaluation of shrinkage with bulk heating and cooling of concrete

There is a risk of cracks occurring up to several days' duration post concreting, dependent on the concrete cooling time. This duration should be considered at an early age of concrete, as it depends on the size of the concrete element concerned. Where:

– $h_0$ (mean radius) <= 150 mm, t = 28 days (concrete cooling duration days);

– $h_0$>= 500 mm, t = 90 days;

– $h_0$, mean radius of the cross section, = 2 $A_c$/ u

– $A_c$ is the concrete cross sectional tensile zone;

– u is the perimeter of that part of cross section, which is exposed to drying.

Intermediate values may be linearly interpolated using the following expression:

$$t = \frac{62}{350}(h_0 - 150) + 28 \qquad [8.1]$$

The concrete tensile zone $A_c$ may be taken as the thickness of the whole concrete considered element (see Table 6.1). The shrinkage may be estimated from the following equation:

$$\varepsilon_{cs} = 0.5\varepsilon_{ca}(t) + \alpha[0.6(T_{max} - T_{ini}) + T_{ini} - T_{\min}(t)] \qquad [8.2]$$

where $\varepsilon_{cas}$ is the autogenous shrinkage at time t, $\alpha$ is the coefficient of thermal expansion, $T_{max}$ is the maximum temperature reached, $T_{ini}$ is the initial temperature of the concrete at the time of pouring, and $T_{min}$ is the minimum temperature to which the structure was exposed during the period up to time t.

*Equation [8.2] comments*:

– the coefficient 0.50, applied to the autogenous shrinkage, reflects the relaxation of stresses when the shrinkage is restrained;

– the coefficient 0.60, applied to the thermal strain, reflects the fact that compressive stresses are generated in the concrete when the strains are restrained during the heating phase. These strains are taken into account by reducing the effect of the temperature rise using a coefficient. This coefficient also takes into account the relaxation of stresses:

– the daily mean value should be used for the $T_{min}$ value, rather than the actual minimum during a given day, as the daily variations in temperature only affect the outer surface to a depth of 25 to 30 cm;

– where default values for $T_{ini}$ and $T_{min}$ are not available, for example, temperatures of 20°C in summer and 10°C in winter in mainland France [MET 15] may be assumed as default values.

## 8.2. Estimating and limiting crack widths

In order to control or limit crack width, in addition to crack distribution, the potential widths should first be calculated. The approach used is case dependent, with crack widths due to loading (calculated in accordance with standard calculation methods, see MC2010 7.6.4.4), combined with crack widths due to delayed shrinkage strains (early age and drying shrinkage). The total crack width should remain below the limiting value related to the concrete exposure class under consideration. Note, the following section is limited to the contribution of early age concrete strains to crack width. The crack width approach and conditions associated with combining other effects are discussed in section 9.2. As per MC2010, the crack width may be calculated using:

$$w_k = 2l_{s,max}(\epsilon_{sm} - \epsilon_{cm} - \epsilon_{cs})$$

where:

$$l_{s,max} = c + \frac{1}{4}\frac{f_{ctm}}{\tau_{bms}}\frac{\emptyset}{\rho_{s,ef}}.$$

For early age concrete, MC2010 suggests a value for $\tau_{bms} = 1.8 f_{ctm}$

NOTE 8.1.–

– In this equation, $l_{s,max}$ is the length over which the force in the reinforcement bars is transmitted to the concrete rather than the crack spacing;

– the ratio $\tau_{bms}/f_{ctm}$ is assumed to be constant and equal to 1.8 (even when the scale effect is considered).

In the case of thermal effects, the shrinkage $\varepsilon_{cs}$ is calculated using the method given in section 8.1.

## 8.3. Estimating restraints at SLS

### 8.3.1. *Approximate calculation of external restraint*

The approximate calculation of external restraint performed for the following two beam examples is considered to be generally acceptable. Where a detailed calculation is not required, this approach may also be applied to walls.

#### 8.3.1.1. *End restrained element*

When cracking occurs, the stress in the concrete is equal to its tensile strength. The following section describes when the equivalent tie function, between two cracks, is similar to that of an edge restrained element. This takes into account the ratio of the cover over the stress transfer length $l_{s,max}$, which is limited, and the strain, which is almost totally restrained. As a consequence crack widths do not develop significantly further. The strain may be estimated using the following equation given in EC2-3:

$$\epsilon_{sm} - \epsilon_{cm} - \epsilon_{cs} = 0.5\, a_e k_c k f_{ctm}(1 + \frac{1}{a_e \rho})/E_s \qquad [8.3]$$

where the coefficients $k_c$ and k are defined in EC2-1-1 section 7.3.2 as follows:

– k = 1.0 in the case of webs or flanges, where h ≤ 300 mm, and k = 0.65 for webs or flanges where h > 800 mm;

– $k_c$ = 1.0 in direct tension only, as in the case of a tie.

In the case of massive elements, where stabilised cracking is usually not reached, the reduction in $f_{ctm}$ value due to scale effects (see section 2.2) should be considered.

NOTE 8.2.–

The EC2-3 equation above may also be expressed as follows:

$$\epsilon_{sm} - \epsilon_{cm} - \epsilon_{cs} = 0.5\, k_c k \frac{f_{ctm}}{\rho}(1 + \alpha_e \rho)/E_s$$

Or alternatively by using the relationship given in MC2010, section 7.6.4.1.1, equation 7.6-5:

$$\epsilon_{sm} - \epsilon_{cm} - \epsilon_{cs} = 0.5\, k_c k \sigma_{sr}/E_s$$

Using the k and $k_c$ values given above for a thick element ($h_0$> 800 mm), it can be seen that:

$$k_{c.}k = 0.65,$$

and as β= 0.60,

$$k_{c.}k \approx \beta.$$

For short-term aged concrete, the equation may therefore be similar to:

$$\varepsilon_{sm} - \varepsilon_{cm} - \varepsilon_{cs} = 0.5\beta\sigma_{sr}/E_s$$

This implies halving of the concrete stiffening tension (the stiffness due to the contribution of the concrete between cracks) at the equivalent tie stabilized cracking stage, in accordance with the MC2010 approach described in Chapter 4.

#### 8.3.1.2. *Edge restraint on a long wall*

The EC2-3 equation – Annex M, is used in this case as follows:

$$\epsilon_{sm} - \epsilon_{cm} - \epsilon_{cs} = R_{ax}\varepsilon_{free}$$

where $R_{ax}$ is the restraint factor.

### 8.3.2. *Detailed calculation of a restraint on a wall*

This case relates to a wall that is restrained continuously along its lower edge. Due to the interlocking effects, the cracks do not extend over the full height of the wall. As a result of differential temperature changes between the wall core and surface, a similar situation occurs during the cooling phase. In this section, a specific calculation method is proposed that takes this effect into account and is based on the observations made on full-scale structures.

Three effects are considered when calculating the restraint:

1) the degree of restraint varies at different points of the wall in accordance with the slenderness ratio (L/H, where L is the length of the wall and H is its height). As a result, not all the cracks extend over the full height of the wall;

2) the relative stiffness between the restraining base (the foundation or preceding concreting lift) $A_0$ $E_0$ and the restrained wall $A_n$ $E_n$ (see Figure 8.1);

3) the effect of interlocking; as the cracks propagate at the base of the wall, the shear stiffness and the normal stress in the wall both tend to reduce.

The proposed approach is based on a method initially developed by ACI [ACI 07] for estimating cracking in a concrete wall rigidly edge restrained on its base and subject to shrinkage that is "external shrinkage restrained on one side". The approach has also been used in a draft Romanian standard GP-115 [RSI 11] and in the CIRIA C660 guide [CIC 07].

The approach is based on a series of curves, which provide the restraint coefficient of the wall as a function of the distance h to the base and the slenderness ratio L/H.

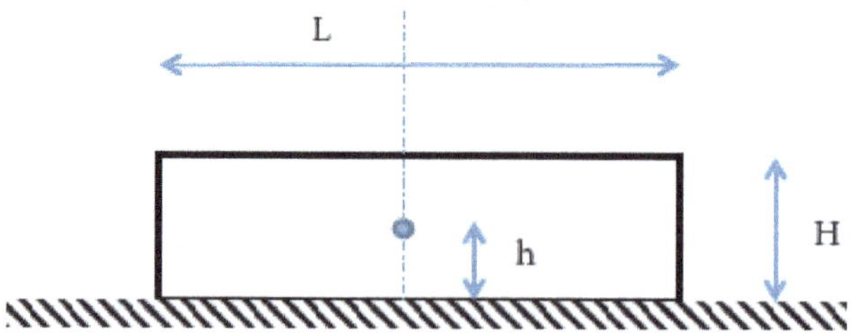


**Figure 8.1.** *Wall edge restrained at its base*

The curves originally given in ACI 207 – 2R have been modified slightly in the CIRIA C660 guide, which suggests the following analytical formulation for these curves:

$$R = \left[1.372\left(\frac{h}{L}\right)^2 - 2.543\left(\frac{h}{L}\right) + 1\right] + 0.044\left[\left(\frac{L}{H}\right) - 1.969\right]\left(\frac{h}{H}\right)^{1.349}$$

where:

– R is the elastic restraint coefficient on a perfectly rigid base;

– h is the distance from the considered point to the base;

– H is the height of the wall;

– L is the length of the wall.

The curves given below show the variations of the restraint coefficient in a wall that is rigidly restrained along its base:

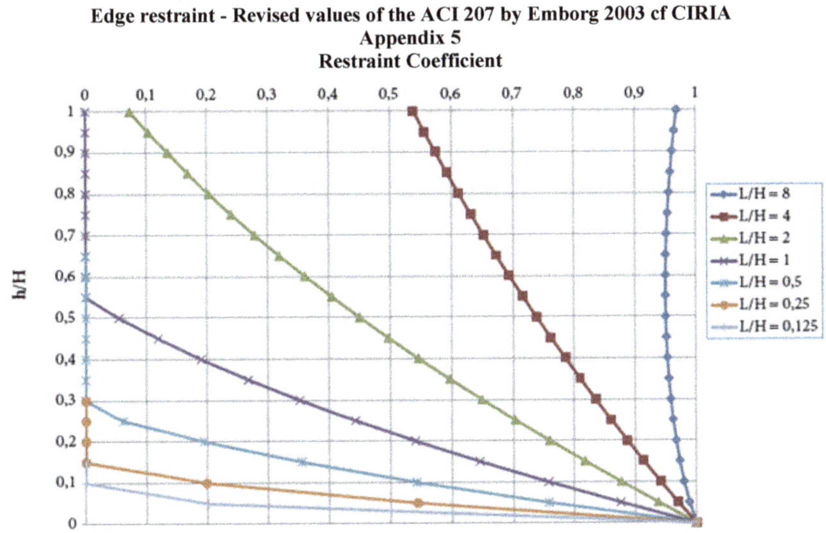


**Figure 8.2.** *Restraint along one edge – revised values from ACI 207 ([EMB 03] CIRIA)*

Figure 8.3 gives an illustration of the analysis results for typical cracking in a concrete element in which shrinkage is restrained at the base.

For long elements (L/H ≈ 10), the curve shows that the stress is uniform throughout the entire height. In the case of short elements (L/H ≈ 2) the curves in Figure 8.2 show that the restraint coefficient tends towards zero at the top of the wall. This implies that cracking across the entire height of a concrete wall cannot occur if L/H < 2.

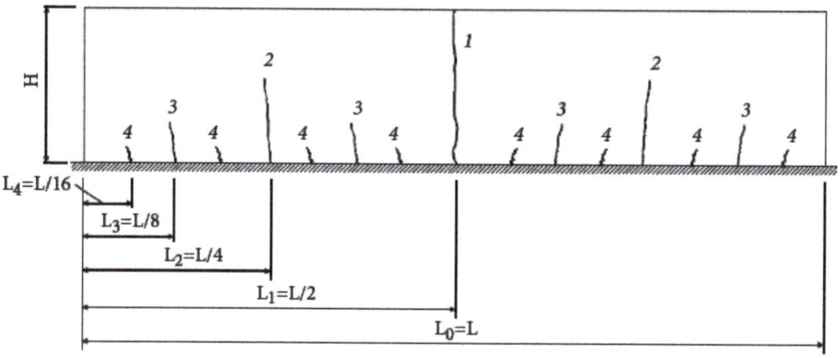


**Figure 8.3.** *Crack formation in a concrete element due to shrinkage restrained at the base*

If the wall has an initial slenderness ratio of L/H = 8, the tension due to the restraint will be mostly uniform throughout the entire height of the wall. After the central crack has formed, the slenderness ratio becomes L/H = 4 and the maximum tension at the top of the wall falls by 50%. After the next cracks have developed, the slenderness ratio is reduced to L/H = 2 and there is no further tension at the top of the wall. When the aspect ratio reaches L/H = 1, the tension only extends as far as 50% of the height of the wall. After the next set of cracks occurs, the slenderness ratio is reduced to L/H = 0.5 and the tension only reaches up to 30% of the height. When the slenderness ratio is reduced to L/H = 0.25, the tension is limited to 15% of the wall height.

The shear stress at the interface between the wall and the base also contributes to the reduction in tensile stress. The Romanian Standards Institute [RSI 11] suggest an equation similar to that of a bimetallic strip, which uses the area of the wall in contact with the surface of the base $L_iB$, the modulus of the base $E_0$, and the stiffness of the wall $A_n$ $E_n$:

$$K_{RO}^{i} = \frac{1}{1+1.05\frac{A_n E_n}{L_i B E_0}}$$

where $A_n$ $E_n$ is the stiffness of the edge-restrained wall (or the new concrete lift).

When the length of the wall L is large, the coefficient is close to 1.0 (base rigidly restrained). As L decreases, the coefficient $K^{i}_{R0}$ decreases. The shear stress between the wall and the base can no longer transmit a significant force over a short length. This coefficient should then be multiplied by the restraint coefficient as a function of the height. Figure 8.4 illustrates this process.

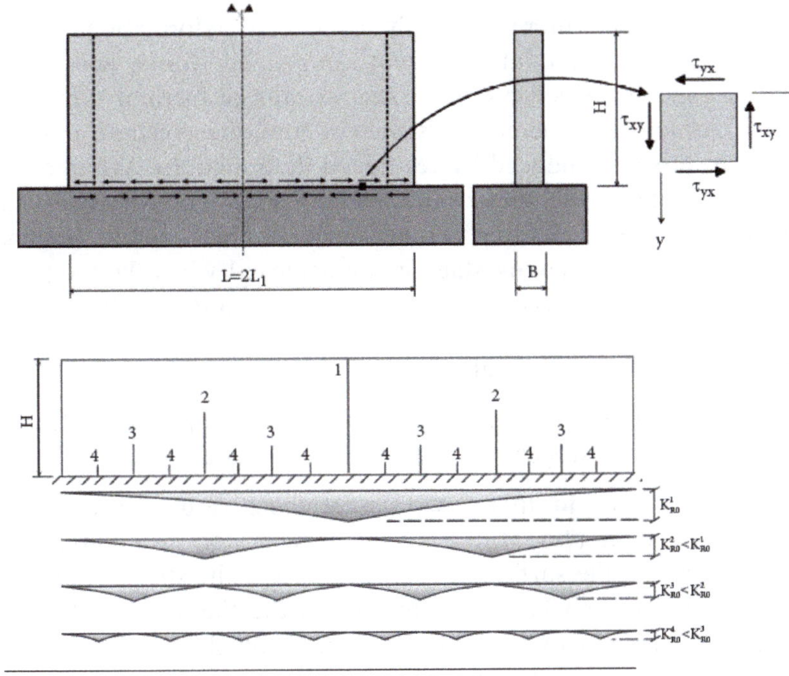


**Figure 8.4.** *The cracking process*

Finally, the following equation can be used to limit the restraint coefficient:

$$R_{bridage} = R \cdot K^{i}_{R0} \quad [8.4]$$

In summary, when $L_i$ / H < 2 the cracks remain limited to the zone in contact with the base. As the distance $L_i$ decreases, the crack propagation becomes more and more limited. The sum of the widths of the various cracks balances the imposed strain. This enables the crack width to be determined according to the distance h to the base. If the crack width is too large, minimum reinforcement as required in MC2010 (section 7.13.5.2) should be used to create new cracks and hence limit the crack widths.

## 8.4. Estimation of stiffness

### 8.4.1. *General comments*

Thermal loads, and more generally restrained deformations, such as shrinkage, creep or differential settlement can generate forces and moments in reinforced concrete structures due to the restraint of thermal expansion or contraction. As tensile concrete resistance is low, the concrete cracks are subject to tensile stress induced by restrained deformations. When concrete cracks, the concrete tensile stresses are relieved and the tensile stress in the reinforcing bars increase, leading to other cracks in the concrete. In general, the stabilised concrete cracking stage is not achieved when the temperature rise is less than 100°C or strains are less than 1000 μ/m [CIC 07].

There are two types of thermal effects:

1) bulk temperature change or global deformation: in this case, the entire structural component is subject to a uniform temperature change;

2) thermal gradient: in this case, the gradient results from different thermal conditions on two of the structure faces, or differential drying between the core and the surface, or between two of the structures surfaces. The thermal gradient can be established when the thermal steady state is achieved at long term or by non-linear transient calculations for short-term scenarios, especially relevant in the case of thick structures.

Thermal stress may be evaluated by considering the stiffness of the member, the rigidity of the section and the degree of restraint of the structure. Structural analysis for concrete structures involves the determination of the reduced member stiffness that arises from concrete cracking and reinforcement yielding.

It would be counterproductive to add further reinforcement to mitigate thermal or similar effects: the additional reinforcement would result in stiffening of the structure, thus increasing the stress due to thermal effects or restrained strains. Therefore, the following methodology consists of verification of the reinforcement already determined by prior calculation.

For example, when subject to thermal gradients, the variation of the thermally induced moment is as represented in Figure 8.5. This figure gives the moment–curvature response for the un-cracked, cracked and reinforcing bar yield parts of the structure. As the restrained thermal curvature is constant, the thermal induced moment decreases as the concrete cracks and the reinforcement yields. The thermally induced moment is dependent on the reinforcement ratio and effective stress, or elongation in the reinforcing bar under mechanical load [BAE 13].

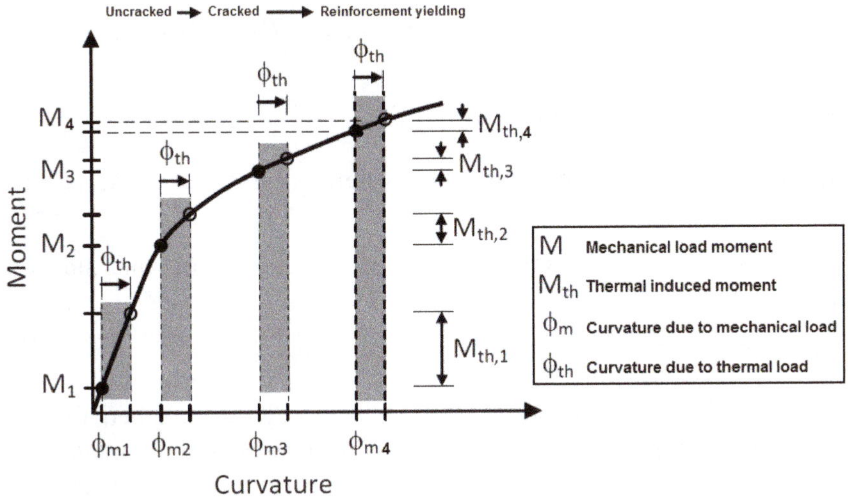


**Figure 8.5.** *From SMIRT-22, curvature–moment diagram*

## 8.4.2. *Simplified method*

Generally during the design phase, a linear finite element model is used which includes all principal, thermal or imposed strains and mechanical, loads considered in the design. The simplified method given in this section assumes that the mechanical forces and bending moments applied to the cross-sectional area, and the effect of the thermal load are derived from a linear elastic calculation.

When thermal or deformation effects are calculated, using linear elastic stress calculations, the induced corresponding loads may be reduced by the following factors. These take into account the cracking of concrete under the effect of heat and are dependent on the linear or non-linear distribution and the normal force.

For reinforced concrete elements with a characteristic compressive resistance in the range:

– 30 MPa $\leq f_{ck} \leq$ 80 MPa and a bending reinforcement ratio $\rho_s \leq 0{,}01$, the reduction factor values are;

– 0.60 for normal operating conditions (SLS.f and SLS.qp);

– 0.50 for exceptional conditions (SLS.characteristic and ULS.fundamental);

– 0.35 for accidental conditions (ULS).

The bending reinforcement ratio $\rho_s$ is defined as equal to $A_{sext}$ / (b • d), where:

– $A_{sext}$ is the cross-sectional area of reinforcement on one side of the concrete surface;

– d is the effective depth of an element cross-section.

NOTE 8.3.– The above reduction factors have been established for special safety requirements, such as waterproofing, air tightness and containment. In accordance with EN 1992-1-1, for standard structures the reduction factors may be considered as zero at ULS.

The reduction factors, due to average deformations, may be applied for thermal gradient or restraint effects. Creep effects and the reduction of concrete modulus, in combination with a rise in temperature, are not taken into account when estimating the reduction factors.

### 8.4.3. *Principles of the detailed method*

The detailed method proposed assumes for internal restraint that:

– subject to uniform temperature variation, being limited to one part of the structure, restrained strain can create a normal force in the locality considered and in the surrounding structural elements. This results in compression in hot elements and tension in cold elements;

– when concrete is not cracked, concrete and steel behave as elastic materials;

– when concrete is cracked, the tensile strength is not taken into account, but the concrete stiffness between cracks is considered using a realistic strain-stress law.

When subject to restrained strain $\varepsilon_0$ effects, the axial force is assumed to be equal to:

$$N = k_{fiss} E_c A_c (\varepsilon - \varepsilon_0)$$

where strain ε is derived from finite element modelling subject to restrained strain $\varepsilon_0$ load, quantity ($\varepsilon$-$\varepsilon_0$) is equal to R.$\varepsilon_0$, where R is the restraint coefficient, either estimated from a linear finite element model or calculated using EC2-3 [NFE 06].

The axial force on a plain concrete section (uncracked) is calculated using the following expression:

$$N_{elast} = E_c A_c (\varepsilon - \varepsilon_0) = E_c A_c R \varepsilon_0$$

It is assumed that when subject to the effects of a restrained strain, only an
axial force is developed in the structural element and there is no strain $\varepsilon_0$moment.

The cracked section calculation is performed by:

– determining the compressed height *x* of the partially compressed section;

– confirming that the section under mechanical load is in equilibrium;

– establishing a stress-strain relationship under a given mechanical moment.

Figure 8.6 provides the corresponding stress–strain state in steel reinforcement bars and concrete when subjected to a force or moment. For each point on the curve, related to the criteria considered (i.e. SLS frequent, quasi permanent or characteristic, ULS), the secant slope of the stress/strain curve provides the stiffness $k_{fiss}E_cA_c$.

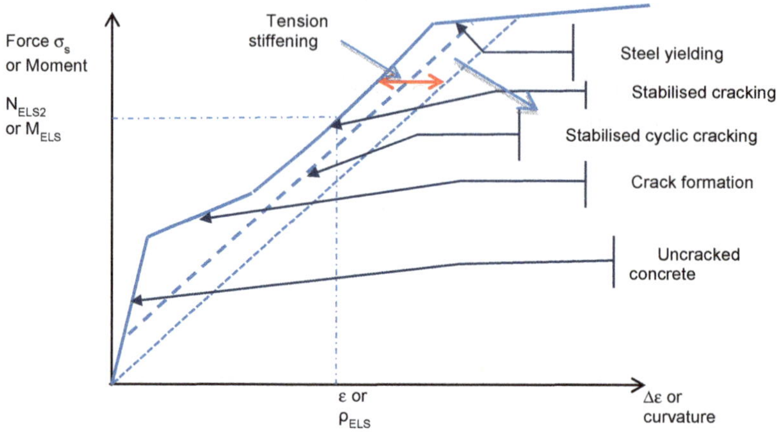


**Figure 8.6.** *Stress-strain diagram; $\rho_{ELS}$ is the curvature value corresponding to the $M_{ELS}$ moment*

When subject to thermal gradient effects, similar to internal restraints, it is assumed that:

– the structural sections (i.e. walls and slabs) do not rotate and create flexural stresses resulting from compression on the hot face and tension on the cold face;

– where the cold face tension remains less than the tensile strength. Where relevant taking account of probabilistic scale effects, that the concrete is not cracked and concrete and steel behave elastically;

– where the concrete is cracked, the tensile strength is not taken into account but the concrete stiffness between cracks is considered using a more “realistic” moment-curvature curve.

When subject to the effect of an imposed strain gradient $\alpha_c \cdot \Delta\theta/h$ or an imposed curvature, assuming that the concrete behaves as an elastic material, the moment when subject to complete restraint is calculated using the following equation:

$$M_{th,el} = \frac{E_c\alpha_c}{1-\nu} \cdot \frac{\Delta\theta}{h} \cdot \frac{h^3}{12} = \frac{E_c\alpha_c}{1-\nu} \cdot \frac{\Delta\theta}{h} \cdot I$$

where:

– I is the mechanical inertia of the plain concrete: $I = h^3/12$;

– $\alpha_c$ is the thermal coefficient of concrete expansion.

When concrete cracks under tension, as a result of the thermal and mechanical stress in the section, the inertia and consequently the moment decreases. Hence the reduction factor applied to the elastic moment in a plain concrete section is:

$$k = \frac{M_{th,eff}}{M_{th,el}} = \frac{12 \cdot I_{fis}}{h^3}$$

It is assumed that under a strain gradient, the moment only applies to the structural element and that there is no normal force resulting from the restrained strain gradient $\alpha_c \cdot \Delta\theta/h$.

The cracked section calculation is performed by:

– determining the compressed height $x$ of the partially compressed section;

– confirming that the section subjected to a normal mechanical forces is in equilibrium;

– establishing a moment–curvature relationship that possibly includes a normal mechanical force.

Figure 8.6 provides the corresponding stress–strain state in steel reinforcement bars and concrete when subjected to a force or moment. For each point on the curve related to the criteria considered (i.e. SLS frequent, quasi permanent or characteristic, ULS), the secant slope of the stress–strain curve provides the stiffness $E_c I_{fis}$.

The equivalent tie stiffness K is used in accordance with the cracking theory described in CEB–FIP model code 1990 [CEB 93] and MC2010 [CEB 12]. Figure 8.6 gives the stress development in a reinforced and prestressed concrete tie and the associated reduction in stiffness.

Additionally, Figure 8.6 highlights the contribution from concrete in tension. An understanding of this contribution is essential to ensure that the thermal loads are not underestimated.

### 8.4.4. *Worked example of a massive element thermal gradient*

#### 8.4.4.1. *Transient thermal gradient calculation*

The preliminary step of the method is to assume that the transient thermal gradient, resulting from thermal shock, can be acceptably modelled using an equivalent bilinear curve.

Two thermal shock cases may be assessed using this method:

1) daily cyclic thermal shock, and;

2) thermal shock with a constant temperature maintained over time.

It is assumed that the thermal shock results from a temperature increase $\Delta T_0$ on one of the faces of a thick element.

When the temperature rise is cyclic (sinus function $\Delta T(0,t) = \Delta T_0 \sin(2\pi t/t_0 - \varphi)$), the temperature at a depth x from the face is calculated using the following
equation:

$$\Delta T(x,t) = \Delta T_0 \cdot e^{-x\sqrt{\frac{\pi \rho c}{\lambda t_0}}} \cdot \sin\left(\frac{2\pi}{t_0}t - \varphi - x\sqrt{\frac{\pi \rho c}{\lambda t_0}}\right)$$

The maximum temperature variation reached at a distance x can be calculated using the following equation:

$$\Delta T_x = \Delta T_0 \cdot e^{-x\sqrt{\frac{\pi \rho c}{\lambda t_0}}} = \Delta T_0 \cdot e^{-x\sqrt{\frac{\pi}{a t_0}}}$$

where time *t* is given in seconds.

NOTE 8.4.– The following thermal characteristics are applicable to a thick element (section 2.1):

$\lambda = 2.3$ W/m °K

c = 1000 J/kg °K

$\rho = 2500$ kg / m$^3$

where the diffusivity $a = \lambda/(\rho c) = 920 \times 10^{-9}$ m²/s and the temperature ingress range is between approximately 0.30 to 0.40 m.

When the temperature rise $\Delta T_0$ at initial time $t_0$ is maintained as a constant, the temperature at depth $x$ from the element surface is calculated using the following equation:

$$\Delta T(x,t) = \Delta T_0 \left[1 - \mathrm{erf}\,(x/(2\sqrt{\frac{\lambda t}{\rho c}})\right] = \Delta T_0\left[1 - \mathrm{erf}\,(x/2\sqrt{((a \cdot t))}\right]$$

where:

– erf(x) is the Gauss error function;

– $t$ is time in seconds.

The temperature propagation may be modelled using a bi-linear curve located at a depth $x_0$ of temperature propagation, so that the heating of this layer, $x_0$ thick, should be equivalent to the actual heating.

For example, the curves in Figure 8.7 give the actual heating related to thermal shock when maintained as a constant and compared to a heating modelled by a bi-linear curve. Figure 8.7 also gives the evolution of $x_0$ according to the duration of exposure to the temperature considered.

#### 8.4.4.2. *Strain and stress calculation*

The following step of the method consists of confirming that the equilibrium of the section, subject to thermal load and mechanical flexural moment, are compatible with the steel reinforcement bar and concrete strains, assuming that:

– element cross-section rotation is limited, which is conservative;

– the strain distribution is constant over the cross section subject to an imposed thermal strain.

The main hypothesis relies on the two following conditions:

– hyperstatic conditions subject to a thermal gradient;

– free deformation conditions subject to an average temperature.

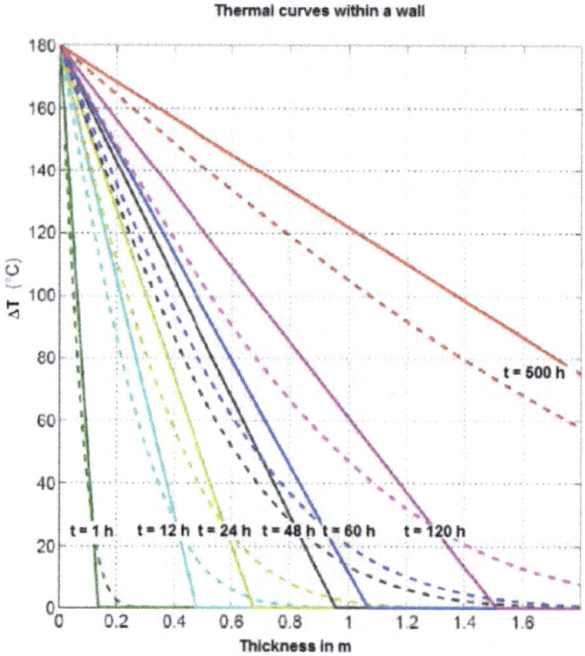


**Dotted curve**: Real evolution of temperature.

**Thick curve**: Equivalent bilinear curve.

**Figure 8.7.** *Example of temperature ingress inside a wall, 1.80 m thick with internal temperature*

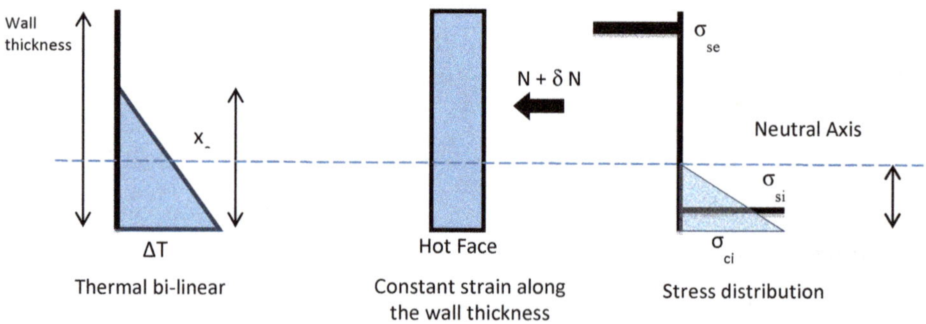


**Figure 8.8.** *Schematic section equilibrium relations under a bi-linear thermal gradient*

Equilibrium equations are provided below, enabling the stress and strain in the materials to be calculated and compared with the design criteria.

– equilibrium of the section:

$$N + \delta N = \frac{\sigma_{ci}}{2}.bx + \sum \sigma_{si}.A_{si}$$

– deformation of concrete hot face:

$$\epsilon_{ci} = -\frac{\sigma_{ci}.(1-\nu)}{E_{cm,th}} + \alpha_c.\delta\theta_{ci}$$

– concrete on the neutral axis:

$$\epsilon_{c0} = \alpha_c.\left(1 - \frac{x}{x_0}\right)\cdot\delta\theta_{ci}$$

– deformation of hot face reinforcement steel:

$$\epsilon_{si} = -\frac{\sigma_{si}}{E_s} + \alpha_s.\delta\theta_{si}$$

– deformation of cold face reinforcement steel:

$$\epsilon_{se} = \frac{N_{se}}{K}$$

– deformation compatibility:

$$\epsilon_{ci} = \epsilon_{c0} = \epsilon_{si} = \epsilon_{se}$$

The mechanical moment, if any, shall be taken into account for all corresponding stresses and strains.

Where:

– $N + \delta N$ is the normal force due to pressure (i.e. thermal and mechanical loads) and excess tension;

– $\sigma_{ci}$, $\varepsilon_{th,ci}$ are the stress and thermal strain in concrete due to the thermal gradient, on the hot face;

– $\sigma_{c0}$, $\varepsilon_{th,c0}$ are the stress and thermal strain in concrete due to the thermal gradient, on the neutral axis;

– $\sigma_{si}$, $\varepsilon_{th,si}$ are the stress and thermal strain in the hot face reinforcement due to the thermal gradient, on the hot face;

– $\sigma_{se}$, $\varepsilon_{th,se}$ are the stress and thermal strain in the cold face reinforcement due to the thermal gradient, on the cold face;

– $\delta\theta_{ci}$ and $\delta\theta_{si}$ are the temperature variations in the hot face concrete wall and at the hot face reinforcement;

– $A_{si}$ is the hot face reinforcement section;

– $\alpha_c$ and $\alpha_s$ are the thermal expansion factors of concrete and steel;

– $E_{cm,th}$ is the elasticity modulus of concrete according to Table 1.4.2.2.

– $E_s$ is the elasticity modulus of steel;

– $N_{se}$ is the axial force in the equivalent tensile tie.

Stresses are positive in compression.

Strains are positive when subject to elongation.

Tensile force in the tie is taken to be positive when subject to tension.

# 9

# Effects of Various Phenomena in Combination

## 9.1. Estimating crack width

Early age cracking can be calculated using the equations given in MC2010, section 7.6.4.4:

$$w_d = 2l_{s,max}(\epsilon_{sm} - \epsilon_{cm} - \epsilon_{cs})$$

where:

$$l_{s,max} = c + \frac{1}{4}\frac{f_{ctm}}{\tau_{bms}}\frac{\emptyset}{\rho_{s,ef}},$$

and $\tau_{bms} = 1.8 f_{ctm}$ as the scale effect on the concrete volume affected by the anchorage and bonding is assumed to be negligible. This is due to bond slip along the reinforcement bars and their associated anchorages.

$$\varepsilon_{sm} - \varepsilon_{cm} - \varepsilon_{cs} = \frac{\sigma_S - \beta.\sigma_{sr}}{Es} + \eta_r \cdot \varepsilon_{sh} \qquad \text{MC2010 Eq.[7.6-5]}$$

where:

– $\varepsilon_{sh}$ is the relative strain due to shrinkage;

– $\eta_r$ is a coefficient for considering the shrinkage contribution (see section 9.2 for its value).

In the case of a massive element when stabilized cracking has:

– not been reached (this is normally the case for cracking due to THM effects at an early age), the scale effects should be taken into account in the value of the mean tensile stress $f_{ctm}$, which is calculated as described in section 2.2;

– been reached, under an applied mechanical force, and where the probabilistic scale effect has not been applied in estimating the stress $f_{ctm}$, then it is recommended, using the CEOS.fr project results for an early age, to reduce the "tension stiffening" $\frac{\beta \sigma_{sr}}{E_s}$ (see Figure 8.6) by multiplying β by a factor of 0.6. The MC2010 [7.6.5] given above, therefore becomes:

$$\varepsilon_{sm} - \varepsilon_{cm} = \frac{\sigma_s - 0.6\beta.\sigma_{sr}}{Es}$$

where:

$$\sigma_{sr} = \frac{f_{ctm}}{\rho_{s,ef}}\left(1 + \alpha_e \rho_{s,ef}\right) \qquad \text{MC2010, section 7.6.4.4.1, Eq. [7.6-6]}$$

When determining the cumulative strain due to different effects, reference should be made to the rules suggested in section 9.2.

## 9.2. Combining effects due to imposed deformations and deformations resulting from in-service loadings

The value applied to the shrinkage coefficient $\eta_r$ used in the above equation is dependent on the following, the:

– restraint coefficients obtained using a method to be defined and the stage of cracking reached in the lifecycle of the element under consideration;

– functional requirements applying to the structure.

For example, in the case of massive elements, where $h_0 \geq 1000$ mm, a value of $\eta_r \approx 0{,}5$ would be acceptable for this type of structure, without any specific water or air leak-tightness requirements, considering a limited drying creep strain resulting from a humidity ratio RH $\geq$ 80%.

A methodology dependent on the functional requirements of the structure is suggested below.

The following approaches are considered separately, namely the cases where:

– a limitation on the crack width is derived directly from structure functional requirements, such as water or air leak-tightness (in particular reference is made to EC 2 Part 3, where the maximum crack widths range between 0.05 and 0.2 mm);

– the verification of the crack widths at SLS is intended to limit the reinforcement bar stresses and ensure the durability of the structure, the surface appearance, and limit any changes to the in-service behavior of the various structure elements due to excessive cracking (with particular reference to EC 2 Parts 1 and 2, where the maximum crack widths are defined according to the type of environment considered, as given in the exposure classes detailed in the application of Table 7.1 N, 7.1 NF, 7.101 N, or 7.101 NF of EC2. These depend on whether the structure is located in France, and is a building or a bridge).

### 9.2.1. *Structures with water or air tightness requirements*

In the case of structures in which the limiting value of crack width is specified to ensure the structure meets its functional and performance requirements for water or air leak-tightness, the irreversible effects of the structures crack widths should be considered to be cumulative. The following effects should therefore be treated cumulatively, the width of:

– cracks occurring at an early age due to thermal–hydro–mechanical (THM) effects accumulating during construction;

– additional cracks resulting from quasi-permanent thermal and mechanical stresses during service;

– additional cracks (if any) associated with the functional requirements applying to the structure for the verification of water or air leak-tightness (considering the value of any characteristic or accidental mechanical and/or thermal and hydric actions). The functional specifications will normally define the combination coefficients and the value of the associated applied forces.

However, in cases where one face of the structure is normally subject to higher levels of humidity, it is proposed that the drying shrinkage effects are ignored, as these will be asymmetric across the wall thickness and hence do not contribute to the transverse crack width.

Where structures are subjected to water or air leak-tightness functional requirements, it is particularly important to consider all possible measures to limit the appearance of potentially transverse cracks at an early age. These measures include controlling the temperature rise and curing time, and selecting an appropriate concrete mix-design, etc. The temperature rise and curing time are controlled through the use of a two step process which includes, alternating the concreting blocks, or using construction joints, enabling concreting in a second stage (i.e. keying together concrete joints between concrete blocks). When these potentially transverse cracks cannot be prevented, their width should, as a minimum, be limited.

Short-term and long-term structural thermal-hydro-mechanical effects may be discounted from the cumulative total effects, if it can be considered that any transverse cracks, likely to open at an early age, will subsequently close permanently. Cracks may close due to various reasons, including:

– pre-stressing closure of cracks at an early age. The CEOS.fr project deliverable 3 [PNC 13] provides a case where cracks that formed at an early age and closed due to pre-stressing, subsequently re-opening during overload tests, exhibiting behaviour identical to the newly formed cracks which were created during these tests – MAEVA [BAR 07] and [CHA 07];

– mechanical flexure. A restrained slab or wall may crack at an early age, due to the structure's restraint, but then be subject to permanent flexure, due to its own weight or ground pressure that subsequently closes the cracks on one side of the structure.

It is generally assumed that the concrete cross-section is:

– entirely in compression, under a combination of loads, with a mean compressive stress $\geq$ 1 MPa;

– subject to flexure, with a mean compressive stress $\geq$ 4 MPa in the compressed part of the section, under the combination of loads considered. Indeed, it is considered that non-stabilized transverse cracking, occurring at an early age due to movement restraint, will not persist, will close cracks, or will not result in any damage that is cumulative with in service cracking.

It should be noted that early age cracking, in some structures required to maintain containment during and after an earthquake, does not have the same profile (i.e. vertical cracking) as the cracking that occurs during and after an earthquake (i.e. inclined cracking in a wall subject to shear stress).

It should be noted that the above mean compressive stress criteria ($\geq$ 1 MPa and $\geq$ 4 MPa) are not leak-tightness criteria. Specific criteria for water and air leak-tightness are specified separately, when required.

### 9.2.2. *Structures with durability requirements*

In most cases, the verification of crack widths at SLS is intended to limit the reinforcement bar stresses, ensure the durability of the structure in the environment to which it is exposed, ensure the surface appearance, and to minimize any changes to the behavior of the structure's various elements, which may result from excessive cracking.

Notwithstanding this approach, it is considered appropriate to perform further verifications using the bounding envelop of crack width values at SLS, without cumulating the effects of each situation.

This implies the assumption, often well-founded, that an initial controlled cracking could have occurred. This produces a relatively homogeneous flexibility in the structure and a degree of adaptation accounted for by a "delayed concrete modulus". This is achieved without affecting the overall function of the structure and the capacity of the reinforcement bars to withstand the applied forces. For example, this is the reasoning underlying the verification of composite steel/concrete bridge decks [KRE 95].

It is still considered necessary that an analysis of the design load cases is performed, ensuring that no permanent forces are omitted from the design. It does not appear that the type of movement restraint affects the choice of which cumulative requirements should be applied.

This approach leads to the following verification steps being used for the cases under consideration:

– *initial verification three days after casting,* is applied to avoid early cracking in the "immature" concrete. When this initial verification is not applied and early cracking is present, there is a high risk of poorly controlled crack widths, due to slippage around the reinforcement bars and the highly viscoelastic nature of the concrete. This initial verification is carried out using the tensile strength and Young's modulus at this early age. The verification consists of checking the crack widths with the material parameters given in section 9.2, under the combined effects of only the thermal and autogenous shrinkage (at three days), the dead load, and the external forces associated with the early stages of construction (three to seven days, including removal of the formwork);

– *verification of crack width control during the construction works, up to the structure entering service*, is an evaluation of the cumulative effects of thermal and autogenous shrinkage, the permanent load, any external mechanical load associated with the construction phase, thermal and hydric forces likely to be exerted in the weeks or months from the start of construction until the structure enters service, and the permanent load that will be applied before the structure enters operation. This verification considers the structure's function in the short term and, as such, the instantaneous modulus of the concrete must be used. However, it should be recognized that there is no "immature" concrete to consider, hence a value of tensile strength for mature concrete ($f_{ctm}$) should be used when sizing the reinforcement. This is required to balance the tensile stress which is likely to be generated during cracking. The CEOS.fr project results [BAR 14] demonstrated that the tensile strength value needs to be weighted by a scale effect factor in order to take account of the concrete tensile zone size ($f_{ctm,\ scale}$);

– *verification of crack width control during the service life*, includes an assessment of the cumulative effects of drying shrinkage, the dead load and the permanent load applied when the structure is in service, together with the effects of the combination of thermal–hydro–mechanical forces representative of the service conditions for which the SLS state is to be verified (e.g. the frequent combination of loads for the design of bridges given in Table 7.101NF of the French national annex to EN 1992-2).

### 9.2.3. *Minimum reinforcement*

The proposed design verification procedure confirms that the reinforcement is sufficient for the following three situations in which cracking should be controlled, under:

– mainly thermal and endogenous effects at a very early age and mechanical effects until formwork removal, at an age when the concrete has not yet achieved its fully matured properties;

– cumulative thermal and mechanical effects during the various stages, from construction to operation, up to an age when non-fragility can be guaranteed in a concrete that has achieved its instantaneous mature properties;

– thermo-hydric-mechanical (THM) mechanical effects in long-term service (without cumulating thermal and hydric effects at an early age), during which the concrete develops its relaxation capabilities.

By taking into account the bounding envelope of these effects, it is the same level of reinforcement which is capable of withstanding the various, potentially independent load cases, as long as the earlier phases have been sufficiently controlled to ensure that they do not result in excessive and irreversible changes to the structure's behavior.

# 10

# Numerical Modeling: a Methodological Approach

## 10.1. Scope

This chapter offers an approach to numerical modeling that takes account of the following non-uniformities within a massive element, namely the:

– temperature;

– drying profile.

The computer-based simulation must also realistically model the changes in the stress field. The domain to which the computer simulation is valid depends on the ability of the mechanical model to demonstrate that the model results lie within acceptable bounds:

– if the mechanical model is capable of modeling the cracking by means of damage theory [MAZ 86], plasticity or any other method, and if the bonding between the reinforcement bars and the concrete is processed in good agreement with the physical phenomenon [ELI 82] then the mechanical simulation may be continued after the first cracking;

– if this is not the case, the non-linear mechanical simulation should be stopped as soon as the first cracking criterion has been reached. Once reached, the stress states, temperature and drying fields can be used, so that the thermal and hydric strains can be considered in a simplified simulation.

## 10.2. Methodology

Numerical models implemented into finite element software packages are capable of providing a direct estimate of the temperature, drying and stress profiles.

These models may also be used to investigate complex superposition of the stress state. This is due to their incremental approach, where the effects of hydration and creep are represented at each time-step. At any time t, the stress may be estimated from the sum of the increments since the initial state:

$$\sigma(t) = \int_{t\ ini}^{t} d\sigma$$

The differential equation governing mechanical behavior of a mass of concrete in a finite element model must be based on a behavior law, expressed in terms of stress increments, in such a way as to take account of changes in the various physical phenomena. This increment between two time-steps in the simulation, must have the following form:

$$d\sigma = S(\zeta)(d\varepsilon - d\varepsilon^{th} - d\varepsilon^{bc} - d\varepsilon^{dc} - d\varepsilon^{sh} - d\varepsilon^{pl})$$

The finite element formulae must be based on this type of behavior law, which is then used in the conservation of momentum equation.

In the above incremental behavior law, the expression S(ζ) is the elastic stiffness matrix for the concrete. This stiffness matrix depends on the Poisson coefficient, the Young's modulus (and the damaged modulus if a damage model is used), and the degree of hydration ζ, which must be calculated in advance using a thermal transient equation as described in section 10.3:

– $d\varepsilon^{th} = \alpha.(d\theta)$ is the thermal strain increment, where dθ is the temperature increment calculated using the thermal transient model, and α is the matrix of coefficients of thermal expansion;

– $d\varepsilon^{bc} = d\varepsilon^{pbc} + d\varepsilon^{rbc}$ is the concrete creep strain increment, which may be broken down into a permanent part ($d\varepsilon^{pbc}$) and a reversible part ($d\varepsilon^{rbc}$);

– $d\varepsilon^{dc}$ is the intrinsic drying creep strain increment. This is only the non-structural part of drying creep since the structural part is implicitly taken into account via the three-dimensional mesh of the structure. The drying creep is a function of the drying increment during the time-step and the level of stress applied;

– $d\varepsilon^{sh}$ is the shrinkage strain increment. It may be split into two components: $d\varepsilon^{ash}$, the autogenous shrinkage resulting from a combination of the water consumption and the negative volume change due to the hydration reactions, and $d\varepsilon^{dsh}$ the drying shrinkage strain increment resulting from a calculation involving the external moisture transfer;

– $d\varepsilon^{pl}$ is the plastic strain including all other sources of permanent strains. This parameter depends on the complexity of the non-linear model used. If this parameter is present in a non-linear model, it usually includes dilatancies resulting from the interlocking of crack lips and slippage between crack lips. It may also include settlement phenomena not included in the shrinkage and creep models.

When an approximate non-linear computer based simulation is used, the transient thermal, transient hydric and non-linear mechanical problems may be coupled. The principles to be followed when implementing each step will be described below.

## 10.3. Thermal and hydration effects

The first step in the process is to carry out the transient thermal simulation, which includes the hydration calculation. The aim of this simulation is to solve the heat propagation problem with a source corresponding to the heat generated by the exothermic hydration reactions. This simulation begins the moment the concrete element is poured, with the reference temperature taken at the moment of pouring. The simulation proceeds until the temperature stabilizes, or until the end of the time period considered, if the system is affected by external thermal phenomena. The conservation law used is that for the conservation of heat (i.e. the thermal formulae used in the finite element software). Including Fourier's Law in the heat conservation equation gives:

$$\rho_C \frac{\partial \theta}{\partial t} = div(\lambda . grad(\theta)) + Q_\infty \frac{\partial \zeta}{\partial t}$$

with the initial condition given by: $\theta(t = 0) = \theta_{ref\text{-}m}$.

In this equation, $\rho_C$ is the specific heat per unit volume, $\lambda$ is the thermal conductivity, and $Q_\infty$ is the heat of hydration per unit volume of the concrete. The hydration rate depends on the temperature and the degree of hydration $\zeta$ already reached as given in the mean chemical affinity equation $\tilde{A}(\zeta)$:

$$\frac{\partial \zeta}{\partial t} = exp\left(-\frac{Ea}{R}\left(\frac{1}{\theta} - \frac{1}{\theta_{ref}}\right)\right).\tilde{A}(\zeta)$$

In this equation, $\theta$ is the temperature in Kelvin. The chemical affinity *A(ζ)* may be obtained directly from either an adiabatic or isothermal calorimetry test. R = 8.31 J/mol is the thermodynamic constant, and *Ea* is the mean activation energy for the hydration reactions. This may be determined from a series of isothermal calorimetry tests at different temperatures or from data given in the literature [BUF 11]. The chemical affinity may also be approximated by:

$$\tilde{A}(\zeta) = \frac{1}{\tau_{ref}}[\zeta^n(\zeta_\infty - \zeta)^m]\left[\frac{1}{n^n m^m}\left(\frac{n+m}{\zeta_\infty}\right)^{n+m}\right]$$

In this equation, $\zeta_\infty$ is the maximum degree of hydration that could be reached. This depends on the type of cement and the water/cement (w/c) ratio. [BUF 11]:

$$\zeta_\infty = 1 - \exp\left(-3.3\frac{w}{c}\right)$$

The parameters m and n are used to adjust the model in accordance with the calorimetry tests. These parameters do not depend on the type of cement. The chemical affinity can be expressed in other forms, particularly for composite binders [KOL 12].

The boundary conditions are those relating to the heat conservation equation. They may be either Neumann type (imposed thermal flux) or Dirichlet type (imposed temperature). It is both realistic and convenient to make use of a surface convection model when simulating the hydration. This convection is combined with the massive element thermal model in order to simulate thermal exchanges with the surrounding environment:

$$-\lambda.grad(\theta).n|_{\text{concrete surface}} = h.(\theta - \theta_\infty)|_{\text{external concrete interface}}$$

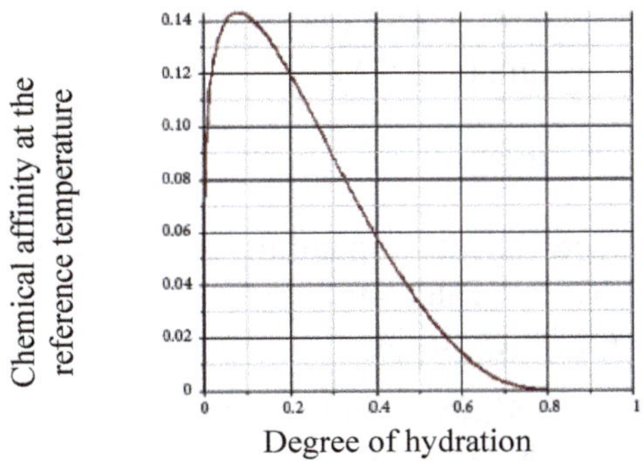


**Figure 10.1.** *Typical chemical affinity (m = 2.2, n = 0.25, $\tau_{ref}$ = 7 hours, $\zeta_\infty$ = 0.8)*

In this equation, *n* is the normal vector exiting from the mass of concrete, and *h* is the surface convection coefficient. This coefficient incorporates the characteristics of the formwork, air velocity, and a simplified form of the radiative flux [BUF 11] $\theta_\infty$ is the external temperature, usually as a function of time.

NOTE 10.1.– it is considered convenient to vary the coefficient *h* in order to take account of the formwork removal.

## 10.4. Drying

The second step in the process is to simulate the drying over a long period, to determine the water content of the concrete element as a function of time. The calculation involves verification of the water mass conservation equation:

$$\frac{\partial w}{\partial t} = div(D_w.grad(w)) - w_{\zeta\infty}\frac{\partial \zeta}{\partial t}$$

where $\zeta$ is the degree of hydration and with the initial condition given by: $w(t = 0) = w_0$

In this equation, w is the liquid water content of the concrete, and $w_{\zeta\infty}$ is the quantity of water required to hydrate all the cement. $D_w$ is the overall hydric diffusion coefficient. This coefficient depends on the permeability of the concrete, the water vapor diffusion coefficient in the concrete, and the water content of the concrete. A possible approximation for this coefficient is given by the Mensi formula [MEN 88]:

$$D_w = A_w.\exp{(B_w.w)}$$

where $A_w$ and $B_w$ are coefficients depending on the quality of the concrete. These coefficients may be calculated by applying an inverse analysis technique to the concrete block drying test results.

The drying boundary conditions may be either Neumann type or Dirichlet type, as is the case for the transient thermal simulation. However, it is preferable to use a surface convection model to simulate the exchanges between the concrete and its environment. The hydricflux condition at the surface may then be written as follows:

$$-D_w.grad(w).n|_{concrete\ surface} = h_w.(HR_{eq}(w) - HR_\infty)\big|_{external\ concrete\ interface}$$

where n is the vector normal to the wall of the concrete, and $h_w$ is the surface hydric exchange coefficient [BEN 05], which depends mainly on the air velocity. $HR_{eq}(w)$ is the equilibrium relative humidity for the water content w, and $HR_\infty$ is the relative humidity of the surrounding environment, which may vary as a function of time. The relationship between the equilibrium relative humidity and the water content depends on the type of concrete. It is usual to make use of a Van Genuchten capillary pressure model [VAN 85] when simulating this property. Ignoring the gas pressure relatively to the hydric depressurization [BAR 99] this becomes:

$$\frac{\rho_w.R.\theta}{M_w} log(HR) = P_w = -\mathrm{M_{sh}}\left(\mathrm{S_r}^{-\frac{1}{\mathrm{mw}}} - 1\right)^{1-\mathrm{mw}}$$

where Sr is the degree of saturation of the concrete calculated as a function of the water content obtained from the drying equation.

$$Sr = \frac{w}{\phi}$$

In this equation, θ is the temperature in Kelvin, R = 8.31 J/mol, $\rho_w$, and $M_w$ are the density and molar mass of the water respectively, $\phi$ is the porosity of the concrete, $M_{ch}$ and *mw* are the calibration coefficients from the water retention curve in the Van Genuchten model.

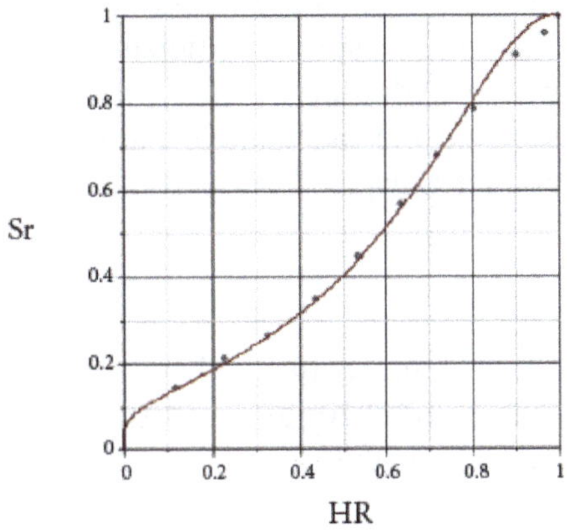


**Figure 10.2**. *Example of a hydric desorption isotherm in concrete (mw = 0.5, $M_{sh}$ = 41 MPa)*

NOTE 10.2.– The conservation equation for the mass of water has the same form as the heat conservation equation. It is therefore possible to simulate drying by means of an thermal analogy in which the temperature is replaced by the water content, the thermal conductivity is replaced by the moisture diffusion coefficient, the heat source is replaced by the water consumption in hydration, and the specific heat is replaced by the porosity of the concrete.

NOTE 10.3.– At high temperatures, the hydric fixation isotherm should be modified (reduction in the adsorption capacity at a given relative humidity Hr as the temperature rises).

## 10.5. Mechanics

### 10.5.1. *Hydration*

The third and final step in the process is the transient non-linear mechanical calculation. This is the last step carried out, as it refers to the solutions of the first two simulations as input data. The published literature contains a large number of models for each of the mechanical phenomena involved in the incremental behavior law. Hence, the following approach is limited to the basic principles associated with the modeling of these phenomena.

The coupling between the hydration and the mechanical behavior is achieved by including a Poisson coefficient and Young's Modulus, dependent on the degree of hydration $\zeta$ in the stiffness matrix. The law most commonly used to associate the mechanical parameters with the degree of hydration is the De Schutter Law, as described in [BUF 11]:

$$\frac{X}{X_\infty} = \left(\frac{\zeta-\zeta_0}{1-\zeta_0}\right)^n$$

In this equation, X is a mechanical characteristic, $\zeta$ is the degree of hydration calculated during the thermal simulation, and $\zeta_0$ is the mechanical percolation threshold at which the concrete passes from a fluid to a solid state.

Exponent values "n" for use in the above De Schutter Law for various mechanical properties are given as an indication in Table 10.1:

| | X | N |
|---|---|---|
| Compressive strength | $f_{ctm}$ | 1.00 |
| Young's Modulus | E | 0.67 |
| Tensile strength | $f_{ctm}$ | 0.67 |
| Poisson coefficient | $1-2\upsilon_\infty$ | 0.67 |
| Crack energy | Gf | 0.5 |
| Bonding stress | $\tau_s$ | 0.5 |

**Table 10.1.** *Exponents for the DeSchutter relationship*

### 10.5.2. *Permanent basic creep*

From a numerical modeling context, there are advantages in using a differential form when modeling the basic creep. For this purpose, several formulae based on Kelvin and Maxwell visco-elastic models are available in the literature [BAZ 01] and [BEN 05]. In particular, [SEL 09a, SEL 16] propose a differential form for the permanent basic creep whose analytical solution is similar to the logarithmic creep function given in

MC2010. In this formula, the increment in the permanent basic creep may be expressed in the form:

$$d\varepsilon^{pbc} = \frac{\varepsilon^{E}}{\tau^{M} C^{c}} dt$$

Where $\varepsilon^{E}$ is the instantaneous elastic strain, and $\tau^{M}$ is a time constant associated with the permanent creep that depends on the degree of saturation *Sr*. The drier the material, the less the creep. It is possible to express this relationship in the following form:

$$\tau^{M} = \frac{\tau^{M}_{(Sr=1)}}{Sr}$$

where $C^{c}$ is a consolidation coefficient used to increase the characteristic time as the material consolidates:

$$C^{c} = \left(\frac{\varepsilon^{pbc}}{\varepsilon^{kbc}}\right) exp\left(\frac{\varepsilon^{pbc}}{\varepsilon^{kbc}}\left(\frac{\varepsilon^{r}}{\varepsilon^{E}}\right)^{nbc}\right)$$

In this expression, $\varepsilon^{kbc}$ is the basic creep potential of the concrete under the stress creating the elastic strain $\varepsilon^{r}$, and *nbc* is the exponent defining the non-linearity of the relationship between the creep velocity and the loading rate. Model calibration of the model is facilitated by the analytical solution proposed for the basic creep test. The associated creep function is then:

$$\phi^{pbc} = k^{pbc^{*}} ln\left(1 + \frac{t}{\tau^{pbc}}\right)$$

where $\phi^{pbc}$ is the creep coefficient:

$$\phi^{pbc} = \frac{\varepsilon^{pbc}}{\varepsilon^{E}}$$

$k^{pbc*}$ is the parameter controlling the amplitude of the permanent basic creep as a function of the intrinsic basic creep capacity $k^{pbc}$ and the level of loading.

$$k^{pbc^{*}} = \left[k^{pbc}\left(\frac{\varepsilon^{E}}{\varepsilon^{r}}\right)^{nbc-1}\right]$$

$$k^{pbc} = \frac{\varepsilon^{kbc}}{\varepsilon^{r}}$$

NOTE 10.4.– The k parameters given in the above expression may be calibrated by applying a least-squares method, to either the experimental results or another form of the creep function. Figure 10.3 gives an indication of the physical significance of the parameters used in the proposed numerical model. It can be seen that the model results, in the long term, are of a logarithmic form, which is in agreement with the relationship given in MC2010. A graphical method may also be used to calibrate these parameters (see Figure 10.3(b)):

### 10.5.3. *Reversible basic creep*

The reversible basic creep may be modeled via a Kelvin modulus, using a characteristic time $\tau^{rbc}$ and the instantaneous elastic strain $\varepsilon^{E}$ to estimate the reversible basic creep increment $d\varepsilon^{rbc}$:

$$d\varepsilon^{rbc} = \frac{1}{\tau^{rbc}}\left(\frac{E}{E^{rbc}}\varepsilon^{E} - \varepsilon^{rbc}\right)dt$$

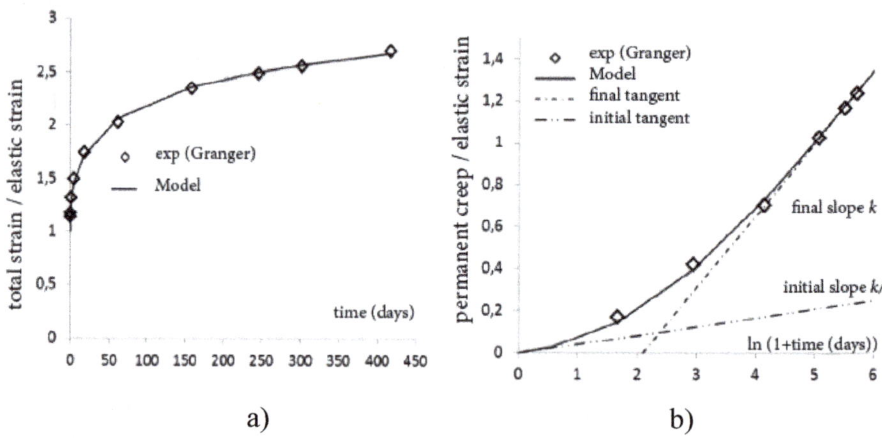


**Figure 10.3.** *Example of calibration of the basic creep function for numerical models*

In this equation: $\tau^{rbc}$ is the characteristic time for the reversible basic creep, as measured during the unloading phase of a traditional creep test (see Figure 10.4), and $E^{rbc}$ is the elastic modulus associated with the reversible creep strain.

NOTE 10.5.– The proposed function used to model the basic creep may be calibrated by reading directly from the experimental creep curve.

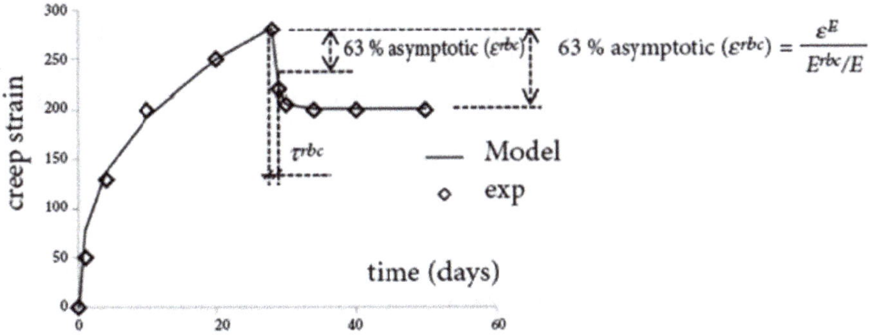


**Figure 10.4.** *Physical significance of the parameters in the proposed reversible basic creep function*

NOTE 10.6.– [SEL 16] proposes an approximation in which the hydration level does not affect the characteristic creep times. This does not mean that the creep is insensitive to the hydration level as, in the proposed model, the creep is a function of the instantaneous elastic strain, which, in turn, is affected by the hydration level through the stiffness matrix in the incremental law.

### 10.5.4. *Influence of temperature on the creep velocity*

The temperature affects the creep velocity. To a first approximation, the characteristic creep time must be reduced as the temperature rises where the reduction coefficient is given by:

$$C^{th} = \frac{\tau^{pbc}}{\tau_{\text{ref}}^{pbd}} = \frac{\tau^{rbc}}{\tau_{\text{ref}}^{rbd}} = \exp\left(\frac{\mathrm{E_W}}{\mathrm{R}}\left(\frac{1}{\theta} - \frac{1}{\theta_{\text{ref}}}\right)\right)$$

In this equation, $\tau_{ref}^{pbd}$ is the characteristic creep time, as defined above, at the temperature $\theta_{\text{ref}}$. This is usually the internal temperature at the time of the creep test. $E_w$ is the activation energy for the creep mechanisms [LAD 10] and R is the thermodynamic constant = 8.31 J/K.

### 10.5.5. *Shrinkage*

The shrinkage strain increment $d\varepsilon_{sh}$ may be modeled by subjecting the concrete to a hydric loading equivalent to the effects of the negative capillary pressure caused by drying and the consumption of water by hydration. Given that the drying model combines these two water consumption mechanisms, the two shrinkage components should also be combined.

$$S(\zeta)\left(d\varepsilon^{ash} + d\varepsilon^{dsh}\right) \equiv d\sigma^{w}$$

In this equation, $d\sigma^{w}$ is the hydric loading increment resulting from the negative capillary pressure increment, where:

$$d\sigma^{w} = b.\,d(S_r.\,P_w)$$

where Sr is the degree of saturation of the concrete with water, and *dSr* is its increment during the time-step. $P_w$ is the hydric pressure (negative), which is a function of the isotherm as modeled by the Van Genuchten equation (see section 10.4).

### 10.5.6. *Drying creep*

The drying creep may also be modeled in different ways. In the CEOS project, [SEL 12a] suggest that the characteristic irreversible intrinsic creep time should be reduced when the material dries, with the characteristic time reduction coefficient given by:

$$C^{dc} = \frac{\tau^{pbc}}{\tau^{pbc}_{ref}} = \frac{1}{1-\left[\left(\frac{d\sigma^{w}}{dt}\right)^{-}\right].\left[\frac{\tau^{pbc}_{ref}}{\sigma^{kdc}}\right]}$$

In this equation, use of the term $\left(\frac{d\sigma^{w}}{dt}\right)^{-}$ means that only the negative rate of variation in the hydric pressure is to be taken into account, i.e. drying. Otherwise, the time constant remains unchanged. This rate of variation is estimated from the current hydric stress state $\sigma^{w}$ and the peak negative pressure occurring in the material up to the present time, where:

$$\sigma^{w}_{min} = \min\left(\sigma^{w}(t)\ \forall t < t_{present}\right)$$

hence:

$$d\sigma^{w^-} = \langle \sigma^w - \sigma^w_{min} \rangle^-$$

In this equation, the notation $\langle\rangle^-$ indicates the negative part of the expression.

NOTE 10.7.– using this approach, the numerical increments in the intrinsic creep strain and the drying creep strain can be combined without increasing the complexity of the numerical model even in transient conditions of temperature and during hydration:

$$d\varepsilon^{pbc} + d\varepsilon^{dc} = \frac{\varepsilon^E}{\tau^M . C^c . C^{th} . C^{dc}} dt$$

### 10.5.7. *Steel-concrete composite modeling*

Most finite element software packages include elasto-plastic models for reinforcement bars and concrete beams. If the reinforcement bar density allows, it is useful to create a mesh solely for the bars and model them explicitly by means of an elasto-plastic model.

NOTE 10.8.– Modeling the reinforcement bars explicitly, prior to cracking, enables the shrinkage stress, restrained by the reinforcement cage, to be directly obtained. Hence it follows that modeling of the reinforcement bars is necessary, even if only the pre-cracking phase is considered. The CEOS.fr national project has shown that endogenous shrinkage, when restrained by the reinforcement cage, can generate stresses of around 10 to 20% of the tensile strength present at first cracking.

When the numerical analysis is continued after first cracking, it is necessary to address the problem of crack location[1] in the concrete and enable relative slip[2] between the concrete and the reinforcement bars, which occurs at the crack edge.

---

1 The problem of localization may be addressed by the use of either; explicit discrete crack models (interface elements, level set), diffuse cracking in the element (Hillerborgh method), or non-local equivalent cracking (non-local method) [PIJ 87]. In all cases, crack propagation must satisfy the cracking energy dissipation condition Gf.

2 The slippage between the reinforcement bars and concrete may be modeled by means of an interface model between the reinforcement bar and concrete models. This interface must behave according to an elasto-plastic bonding model of the type given in MC2010 or [ELI 82].

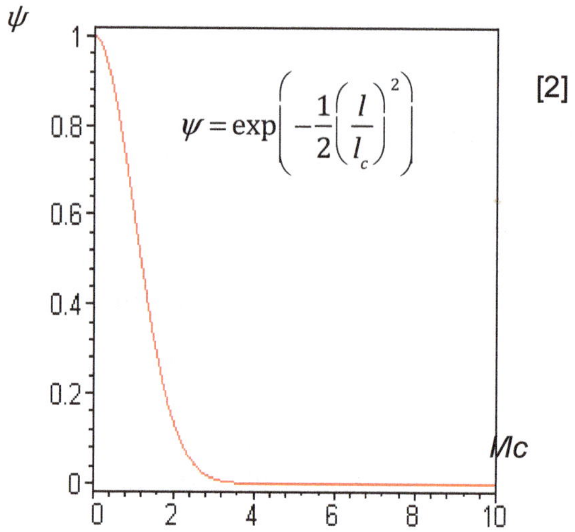


**Figure 10.5.** *Probabilistic weighting function used to weight the Weibull integral*

### 10.5.8. *Statistical scale effect*

The finite element software numerical models provide access to the entire stress field. Through this approach it is possible to use the stress field, at each time-step, to estimate the equivalent volume for applying Weibull's law. Based on the modifications made to Weibull's law in Chapter 2, limiting the sizes of the integration zones, it is possible to make use of the modified Weibull integral, proposed by [SEL 14a, SEL 14b] in the Mefisto project (associated with CEOS.fr project) in order to calculate the equivalent loaded volume:

$$V_{eq(M)} = \alpha \left( \frac{\sigma_{max}}{f_{ctm}^{V_{ref}}} \right)^{-k}$$

where k is the Weibull exponent described in Chapter 2, and $f_{ctm}^{Vref}$ is the mean tensile strength measured from the reference specimen, where:

$$\alpha = \int_V \left( \frac{\sigma_{max}}{f_{ctm}^{V_{ref}}} \right)^{-k} \psi(l/l_c)dV$$

In this equation, the function $\psi$ is used to limit the integration zone, given the maximum distance $l_c = 1.25$ m.

Once the integral has been calculated, the equivalent volume can be estimated and applied, as described in Chapter 2, to estimate the strength value for use in the cracking simulation.

$$\frac{f_{ctm(M)}}{f_{ctm}^{V_{ref}}} = \left(\frac{V_{ref}}{V_{eq(M)}}\right)^{1/k}$$

NOTE 10.9.– The non-local integral used to estimate $\alpha$, and hence $V_{eq}$ , may be calculated numerically, either explicitly using a non-local algorithm, or implicitly via a second gradient method as proposed by [SEL 14a, SEL 14b]. This second gradient method is significantly faster to use, and gives the most probable strength at each point in the element at each time-step in the non-linear simulation.

Other methods may be used to take account of the random nature of the tensile strength of the concrete, particularly the random field's method and the method used to represent the meso-structure explicitly. As these are simulation methods using random sampling, a sufficient number of samples should be taken in order to guarantee that the solutions are representative.

## 10.6. Example simulation

An example of a simulation which makes use of the modeling principles described above is given in the CEOS.fr national project. This example considers RG8b, RG9 and RG10 test blocks subjected to restrained shrinkage, and is described in detail in [BUF 14]. Only the main aspects of the modeling are described below. The mesh used is shown in Figure 10.6.

### 10.6.1. *Thermal and hydration simulation*

A convection model is used for the exchanges between the concrete mass, modeled in 3D, and the external environment. The exchange coefficient takes account of the specific nature of the formwork:

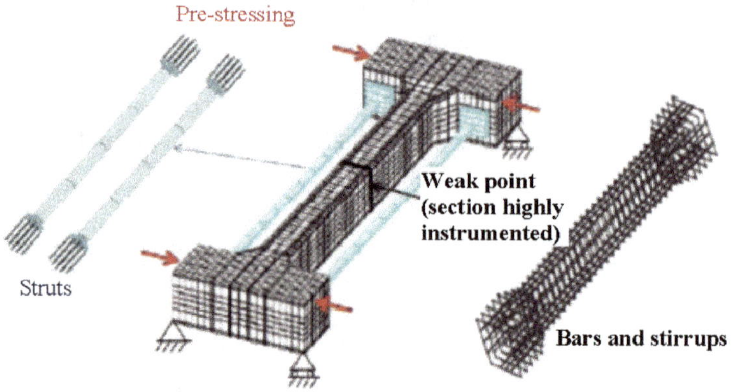


**Figure 10.6.** *Mesh applied to the concrete, reinforcement cage and metal struts in ties subject to restrained shrinkage, taken from the CEOS.fr national project*

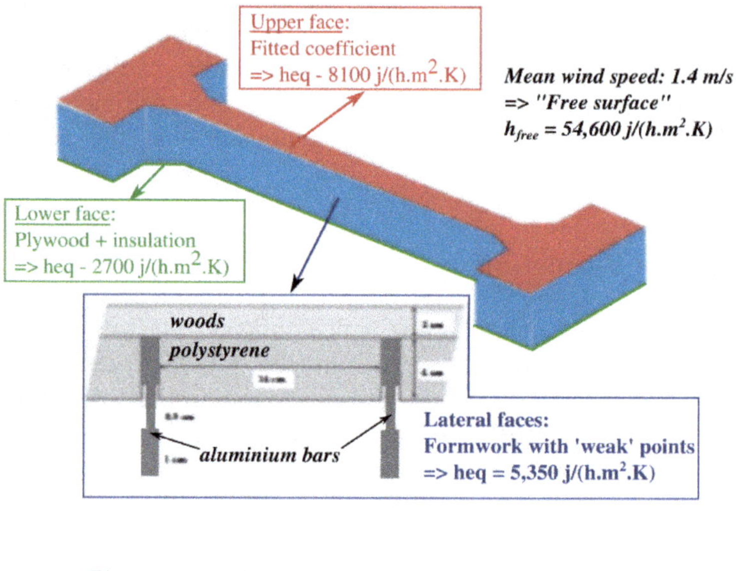


**Figure 10.7.** *Convection models used in the thermal and hydration simulation of a CEOS RG8b block*

The following graph may be used to compare the temperatures at two points on the straight section of the block with experimental results:

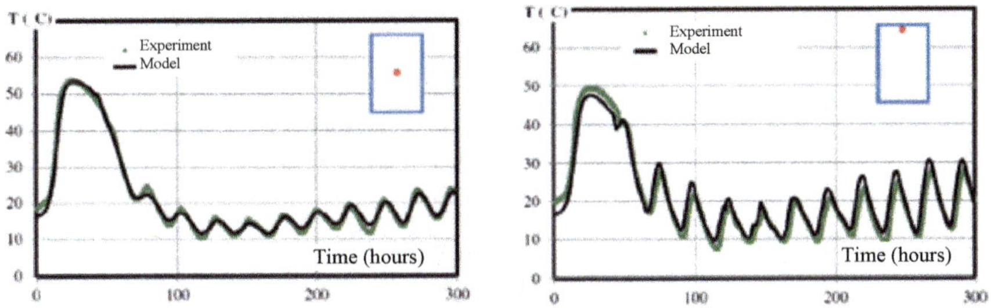


a) Core temperature    b) Upper face temperature

**Figure 10.8.** *Comparison of the numerical and experimental results for RG8b test block*

# 11

# Recommendations for the use of Measurements on Mock-up Test Facilities and Structures

The following key recommendations mainly focus on measurements on massive reinforced concrete structures, which include massive elements. For such structures, normal rules should be adapted with reference to mock-up test facilities developed ahead of construction. These mock-up tests should be representative of the mechanical characteristics and operational rules of the full-size works.

Monitoring programmes for these structures provide information, which require an increased understanding of the on-going performance of aging assets. The instrumentation used in the monitoring programme provides initial quantitative data during the construction phase. This helps avoid the risk of early failure due to excessive stress and damage that normally occurs (i.e. early age behavior). In addition, instrumentation also provides long-term consistent behavioral data records, which enables verification that the structure is performing as expected. Therefore, integrity monitoring of the structure should be a key part of any owner's risk management programme.

Currently, only a few documents provide the methodology which is required for these measurements, either for massive concrete structures or representative mock-ups (test facilities). The aim of this chapter is to summarize the methodology applicable to the measurements. These guidelines benefit from the experience acquired during the CEOS.fr tests and feedback from in-service massive structures.

## 11.1. General methodology of the measurements

Two types of approach are presented in the following paragraphs: the first is applicable to massive concrete structures, the second to large-scale mock-ups. The measurement of massive structures allows the owner to assess the features during the construction phase and lifetime. This approach enables on-going performance assessment of aging assets. The structural measurements should be compared with the theoretical predicted results, which are based on analytical calculations or finite element simulations. Periodic inspections and advanced computational structural analyses are considered to be complementary approaches.

Some values, such as stress, cannot be directly measured. Hence, these measurements are provided through geometrical data, strains on a length basis, or variations from the initial measurement taken. As few sensors provide an absolute measurement datum, it is necessary to take into account the environmental conditions as well as the sensor accuracy.

*For mock-ups*, the objective is to obtain relevant, usable and realistic data which can help evaluate temperature, hydric, strain and stress fields. This helps to mitigate early age failure risks, such as concrete cracking. The use of these mock-ups requires specification of the measurement technique, particularly for instrumented reduced scale structures.

To validate THM phenomena, and/or structural mechanical behavior, the mock-up tests are usually reduced in size (generally 1/3 scale) or have representative elements designed in accordance with controlled limit conditions. The assessment of these tests is generally performed in a laboratory or on a construction site, where the life expectancy of the mock-up will be relatively short compared to the actual full-size structures. Hence, the phenomena considered in these tests are relatively short-term. To support this assessment, a large number of sensors can be used and a cross-comparison of data made so as to focus on the functional performance of the test facilities.

The guidelines below result from the experience feedback of the CEOS.fr project, which considers the measurement methods applied for the set-up of the large scale demonstration test facilities behavior. These include:

– massive beams with free shrinkage during the concrete maturing phase (i.e. CEOS.fr RL blocks), tested in four points of flexion;

– massive beams with restrained shrinkage during the concrete maturing phase (CEOS.fr RG blocks) tested in four points of flexion;

– concrete walls (third scale) with an alternate shear loading applied.

For the massive beams RL and RG, continuous monitoring is applied following casting (i.e. the early age exothermic stage), for the duration of the concrete maturing phase, and during the mechanical bending test activities over a period of a few weeks.

*For concrete structural monitoring*, instrumentation enables verification that the assets are performing as expected and is also used as a diagnostic tool in the identification of structural degradation. Consequently, the location of sensors is carefully chosen and the number of sensors optimized from the beginning of the project.

*For the above*, a minimum set of initial data is required, in particular measurement of temperatures and calibration, using witnessed specimens. These specimens identify the initial conditions of the early age concrete (i.e. combined exothermic phenomena and maturing phase reactions). Appropriate procedures for this approach are detailed in section 11.2.

*For the selection of instrumentation*, two criteria must be considered:

– Qualitative criteria;

- metrological measuring devices have to be qualified in terms of uncertainties. This helps reduce the estimated parameter measurement variability when used in the numerical simulation and structural analysis,

- human factors: operators should be suitably experienced in the equipment used, including in its implementation, adaptation to construction site operating conditions and test instrumentation data signal storage;

– Quantitative criteria;

- reliable and qualified measuring devices shall be numerous and located in predetermined zones, as defined by the numerical simulation,

- acquired measurements shall be taken at an appropriate frequency, when compared to the change of the phenomena, and be adapted to the supporting calculations,

- data processing and maintenance of the numerous test points must be designed to take into account the available technical equipment and expertise/capability of the developer.

Given that the mock-up is the “reference” datum for the structure’s lifetime, providing information which can be used for comparison with the actual full-size structure results, it is recommended that the mock-up is retained for future use. In order to obtain reproducible and comparable data over a long period, the durability of the instrumentation must be considered by the designer.

A glossary specifying the dedicated metrological vocabulary is available (see the French standard NF ISO/CIS GUIDES 2011, International Vocabulary of Metrology – Fundamental and General Concepts and Associated Terms – VIM).

### 11.1.1. *Preliminary general approach*

For an improved understanding of the phenomena, it is recommended that the following are considered:

– completion of concrete suitability tests in the laboratory beforehand, taking into account features that are significant to the scale of the structure, following appropriate review;

– adaptation to the local environmental conditions in the choice of the monitoring technology (i.e. the actual conditions on the construction site and future, in-service operational conditions of the structure);

– assessment of the concrete behavior through suitability tests;

– confirmation that the measuring sensor installation has been performed by a suitably qualified firm using experienced operators in accordance with procedures under an accredited quality assurance system;

– provision of calibration and/or a manufacturers certification, confirming the quality of the equipment, sensors, data acquisition and means of measurement;

– preparation of measurement device, and cabling equipment drawings, along with operational procedures for use prior to and following installation;

– adoption of a formal rigorous document management system;

– confirmation of the monitoring measurement frequency applied and any potential change through the structure's lifetime;

– compatibility of the various data acquisition systems and inspection methods or visual identification methods;

– retention and protection (using the last software updates) of the measurements, results, notes, historic reports of the construction site, all associated documents and pictures on digital platforms (future proofed lifetime records);

– storage of the associated test specimens in an accessible and adequately protected location, such that it is possible to eventually take samples and/or create, through cutting (i.e. sawing) appropriate sections which enable examination of the fracture surface of the most significant cross-sections.

### 11.1.2. *Selection and choice of measuring devices*

Firstly, the chosen instrumentation must take account of the massive concrete element temperatures (T), hygrometry/water ratio (H), which depends on the degree of hydration, and the mechanical strains (M), which are non-uniform in the section considered. Note the three major parameters, THM, used to assess a concrete structure's performance (local or global behavior) are associated and their evolutions interdependent.

The use of suitable instrumentation enables confirmation of the mechanical behavioral requirements of the structure, in terms of its short-term strains, then as the structure matures, the on-going behavioral requirements, in particular cracking and shrinkage.

For massive concrete structures and reinforced concrete sections (representative of blocks, rafts, walls, etc.), two physical phenomena should be investigated and associated measurements made:

– internal phenomena, such as internal thermal gradients and the resulting auto-stress;

– external phenomena, such as global and local strains, crack detection, and environmental measurements which have a direct influence on the behavior of the structure (air temperature, relative humidity of the air, rain, wind, speed and wind direction, period of sunshine, etc.). The measured strains and external effects are uncorrelated. The proposed measurement methodology consists of selecting suitable sensors which are matched to the parameters and the physical amplitude measured;

- T: temperature,

- H: hygrometry (moisture content, relative humidity RH, water ratio and hydric pressure),

- M: mechanical measurements (strain, distortion, crack, displacement, etc.).

### 11.1.3. *Method of measurement selection*

The main measurements considered require the collection of data from internal element sensors. Independent analysis of this uncorrelated data is used to predict the behavior of the structure (strains, stress and cracks). Following this, the data and results collected are used in the numerical formulae given in the codes EC2 and MC2010 (see Chapter 7) and numerical simulation models.

Currently, there is no standard methodology or agreed implementation method which can be applied to measurements and their use as mechanical indicators. As a result, no common approach exists between the structure's owners, which allows or facilitates comparison of structural measurements.

For example, based on the experience gained during the CEOS.fr project, Table 11.1 provides relevant good practice for the selection of sensors.

### 11.1.4. *Measurement data-mining and analysis*

Monitoring programmes provide local data. The structural analysis provides mean deterministic mechanical values such as temperature, strain or stress. As a consequence, the correlation between the results obtained from numerical calculations and those obtained from direct measurements require particular attention in order to:

a) *Estimate the absolute measurement from the relative measurement.* Few sensors give absolute values; for example, a rigorously calibrated sensor, when compared with a calibrated scale, gives a direct absolute value (for example Pt 100 for temperature).

| *Parameter to be measured* | *T or H or M* |
|---|---|
| Inventory of existing measurement systems | Compilation, usually expressed as a builder's balance sheet, of the existing means used for measurement of the THM parameters. |
| Existing assessments and experience feedback where available | Compilation of existing assessments and experience feedback, based on the methods already used to measure THM parameters. This approach supports the optimization of measurement system used. |
| Metrological performance of the selected method | Manufacturer's data sheet specifying the method to be applied for metrological calibration of the measurements system. |
| Conditions of implementation | Produce technical specifications and associated location drawings for the measurement devices. |
| Concrete physical characteristic conditions experienced during the casting process and at an early age. (8-day concrete maturing phase) | T: 8°C to 70°C maximum<br>P: maximum hydrostatic pressure (lift height during casting)<br>M: pressure on the sensor body (lift height during casting and shocks in the pouring of the concrete)<br>C: PH 13.5, liquid concrete |
| Environmental characteristic conditions experienced during the structure's lifetime | T: minimum X°C to maximum Y°C (static or dynamic) temperature<br>P: maximum pressure (static or dynamic local hydrostatic pressure)<br>RH: relative humidity (0% to 100%)<br>M: strains on the sensor body (static or dynamic)<br>C: local conditions for the pH of the structure |
| Vector and distance of transmission | Wired devices: cable length influence and its possible size modification (reduction or extension).<br>Cable robustness when subjected to various environmental conditions.<br>Transmission by radio system: in agreement with the regulations on the current wavelengths and the range between receivers and transmitters. |
| Number of measurements performed | Quantity of measurements acquired and stored.<br>Measurement frequency (static: minute, hour, daily, monthly, annual, or dynamic measurements/second).<br>Life expectancy of the measurement devices required, in support of the objective of a long unit lifetime (often more than 60 years): with multiple and potentially redundant measuring devices. |
| Measurement data-mining | Software, storage and eventual Global Positioning System (GPS). |

**Table 11.1.** *Choice of sensors*

However, numerous sensors, in particular long optical fiber cables, give a relative value which requires comparison with their measurement basis. If a long base sensor is sensitive to other conditions, such as temperature, hygrometry etc. it will be necessary to correct the values for any associated effects. Therefore, it is essential to have reliable temperature measurement (Pt100 sensor/probe) and if possible the moisture ratio of the material.

b) *Detection of the occurrence of the first crack* in the measured media considered as continuous. Detection of the first crack occurring in one direction is the consequence of a medium which presents discontinuities. Therefore, as a result of this first crack, the material section decreases, dependent on the crack size. The crack depth in the section is difficult to assess and it influences directly and proportionally the strains or the displacement when the structure is subject to tension or compression. Following the occurrence of other cracks, the remaining un-cracked section may relax or contract. These relative strains are included in the total strain measurement. Which, as a consequence, makes it difficult to quantify the change in strain following crack initiation, and difficult to detect the occurrence of cracks. Therefore, it is also recommended that other redundant means of crack inspection (global measurement), such as visual inspection or photogrammetry, etc., be used.

c) *Other measurements required to avoid the introduction of disturbances*, which are sometimes introduced by an accumulation of sensors and other measuring devices within the test specimen, especially at critical locations in the structures casting.

## 11.2. Mock-up measurements

For the CEOS.fr project experiments, an expert group selected the most appropriate cost-effective instrumentation, taking into account the anticipated amplitude of the phenomena, based on the engineering code (EC2 and/or MC2010) predictions, as confirmed by the numerical simulations and cement material formulation suitability tests. The main objective of the measurements is to focus on a better understanding of the local phenomena of damage and cracking with time, plus the effects of environmental and other externally imposed events.

Section 11.5 provides an example of the CEOS.fr project experience for the measurements implemented, solutions adopted and their associated justification.

Based on the use of monitoring in the interpretation of phenomena acting on test specimens, and using results from other existing structures and test facilities, recommendations for measurements applied to early age concrete and during mechanical loading were established. Finally, as part of research into a specific phenomenon, or a specific structure function, requiring the use of test facilities, a full analysis of the data using finite element simulation shall be performed. This helps to refine the design of the mock-up. Once the mock-up has been constructed, the actual boundary conditions shall be corrected to take any measurement uncertainties into account. This approach is essential for ensuring that the use of the mock-up is able to repeat realistically the physical processes occurring in the works.

### 11.2.1. *Measurement of parameters*

The measurement of parameters has to take into account the phenomena acting on the concrete, recognizing that the three THM parameters considered are connected, time dependent (especially for early age concrete), and are strongly linked with the concrete performance and physical behavior. Any correction associated with the phenomena acting on the concrete must consider the external ambient thermal and hydric conditions. For these three THM parameters, the evolution of the test specimen strain and cracking (crack spacing and widths) should be taken into account.

#### 11.2.1.1. *Temperature*

Given the influence of temperature on the concrete mechanical characteristics (i.e. too great an internal auto-stress leads to internal cracking) and physico-chemical characteristics and their evolution over time, the measurement of early age concrete temperature is key. Therefore, in order to obtain the internal gradient of the massive reinforced concrete structure, the concrete core temperatures should be measured accurately, reliably and from a number of locations. Use of the least intrusive possible sensors, which give an absolute temperature value, is recommended (Pt100 type). This supports the collection of accessible, reliable and absolute temperature measurements.

Hence, if standards imposed by manufacturers are respected, this supports the reliability of the measurement. For example, tolerances for a given temperature measurement T°C = ± (0.15°C + 0.002.T°C) - Standards IEC 751 (1983), BS on 1904 (1984) and DIN 43760 (1980) (NF 60584.1: on 1996 or the CIS 584.1: on 1995). The use of Vibrating Wire Strain Gauges (VWSG) enables local measurement of temperature and micro-strain from the same location. Once the accuracy of the temperature measurement, by comparison with Pt100 sensor/probe outputs, has been established, the local micro-strain measurement provides the local shrinkage of the concrete (see section 11.2.1.4).

#### 11.2.1.1.1. Measuring device location, distribution of temperature probes and distributed optical fiber temperature measurement

In order to avoid an internal sulfatic reaction in concrete (RSI–DEF), it is recommended that the maximum concrete temperature should not exceed 70°C (see section 2.1 Note). Notwithstanding the adherence with recommended practice, the special structure reinforced concrete thickness can result in a local core temperature which exceeds the recommended 70°C limit. This can lead to significant thermal temperature gradients between the structure's core and surface, or between two surfaces of the same structure.

As a minimum order to obtain reliable measurements, three temperature sensors should be incorporated in the mock-up cross-section, both width and height, repeated each meter length of the mock-up (Figure 11.1).

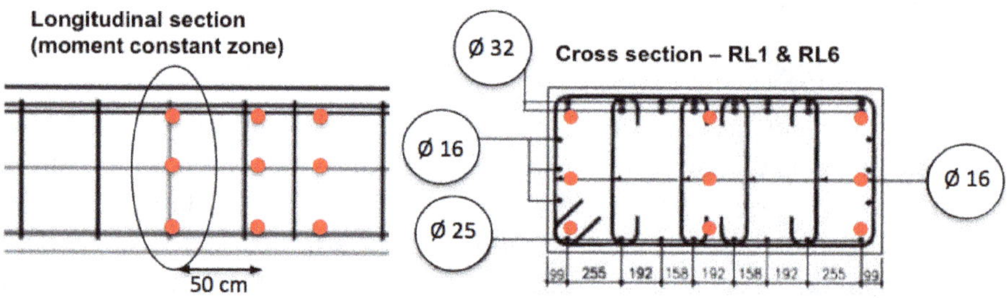


**Figure 11.1**. *Example of optimum location of temperature probes in the case of CEOS RL1 tested beams*

#### 11.2.1.1.2. Temperature measurement before formwork removal

Based on CEOS.fr project feedback, it is considered advisable to check, before the formwork removal, that the:

1) concrete temperature differential $T_{surface} - T_{ambient}$ is limited to a value based on the concrete formulae and type of structure element considered (beam, slab, etc.);

2) concrete internal temperature gradient never locally exceeds a value, which corresponds to the maximum tensile strain concrete capability; this

temperature gradient can be determined using a model of the structure (from a minimum of third scale to full scale).

The above checks shall be applied respectively, taking into account that check 1 must be performed on the construction site, so the period of formwork removal can be optimized in accordance with the temperature differential $T_{surface} - T_{ambient}$.

#### 11.2.1.1.3. Temperature measurement at formwork removal

In order to avoid thermal shocks, it is necessary to ensure that the removal of formwork occurs with a limited risk of significant thermal gradients between the element surface and core, or between element surfaces. This, along with confirmation of the exothermic peak temperatures for all zones (even in the hotter zones) and that the internal temperatures are stabilized at a value defined in accordance with the external temperature, helps to avoid any thermal shock.

Note that when removing formwork at an early age, the concrete has not yet reached its fully matured state, or hence its expected modulus. The lower the concrete modulus is, the higher is the risk of concrete cracking.

EXAMPLE.– CEOS.fr concrete RL beam tests.

In the case of CEOS.fr RL beams, surface micro-cracking was identified as soon as the internal/external differential temperature exceeded 40°C.

The thermal expansion coefficient of the CEOS.fr project concrete is approximately 120 µm/m/10°C (where the concrete coefficient of expansion: $\alpha$ = 12 µm/m/°C). For this concrete, a significant temperature differential, measured over a small length, could lead to internal stresses and damage. In the case of the concrete used for the CEOS.fr project test beams RL1 and RL6 (see section 1.1), based on $f_{ctm}$ = 4.67 MPa, and taking into account the scale effect, a maximum temperature gradient of 10°C/0.5 m is considered acceptable. Cracks occurred in the following two configurations:

– RL6 beam, where an exothermic core temperature of 76°C was reached. This resulted from the combination of external conditions, imperfect formwork insulation, and a maximum concrete surface temperature of 60°C on the beam's lateral faces, even when the core temperature was still increasing. In addition, the temperature of the lower portion of the concrete beam (isolated from the site soil) followed the core temperature. As a result, the temperature differential of 10°C between the lateral faces and lower face of the beam was reached over a length less than 0.5 m; and

– RL1 beam, where an exothermic peak core temperature of 60°C was reached. In addition, removal of the formwork was performed at 48 hours precisely. Upon removal at this time, the concrete temperatures were still significant (i.e. 55°C at the core), the ambient temperature was 13°C and sudden cooling of the surface lead to a significant change between the core and surface temperatures. As a result internal cracks occurred in this configuration.

#### 11.2.1.2. *Moisture content*

Concrete, as a cementitious material, is a porous structure that exhibits a permeability ratio whose variability is dependent on the concrete mix formulae (cement type and additions, nature and aggregate size, water ratio and additives), casting process (direct casting, pumping, vibration), environmental conditions during the maturing phase, the degree of hydration and concrete layer dimensions. As the cement structure loses water during the maturing phase, resulting from an endogenous chemical reaction, water continues to be absorbed or released over time (i.e. rehydration or dehydration phenomena).

The water content of concrete is one of the criteria which control its mechanical behavior, including shrinkage (modification of the elasticity modulus) and creep.

To support the functional capability and sustainability of civil works, it is necessary to understand the hygrometry changes in the in-service structure. Strain data obtained from measurements on the structure are de-correlated from hygrometry changes provided by the predictive numerical simulations and the laboratory test results performed on samples using the same concrete formulae.

Concrete water content is hence a major parameter to be measured. This measurement is crucial during the concrete maturing phase given the influence of hygrometry on the concrete expansion coefficient and elasticity modulus. Following the early age phase, which is generally at least 28 days[1], hygrometry monitoring provides a practical indication which helps to quantify, in the long term, the global concrete strain and the hydric shrinkage part of the concrete mechanical creep[2].

Other measurement systems, for example TDR type, which measure reflections or frequency waves, or "Pulse" type measurement provides in-service humidity ratios. However these measurement systems still need to be qualified and, where necessary, improved.

To ensure that the concrete formulae performance criteria are met (permeability, modulus, shrinkage, etc.), and to include additional measurement which separates the thermo-mechanical coupling from hydric conditions, it is necessary to compare in-service measurements with standardized laboratory test measurements. Consequently, temperature sensors are prescribed in addition to strain extensometers, used in three directions to establish the thermo-mechanical behavior of the instrumented mock-up.

The calibrations of the sensors and comparison with reference standards are a major issue.

#### 11.2.1.3. *The evolution of concrete elasticity modulus (Young's Modulus $E_c$)*

The Young's modulus is established experimentally on test specimens at a duration of 3, 7 and 28 days in accordance with NF EN 12390-13, the static elasticity modulus determined in compression. To determine the elastic modulus, the tests apply three load cycles at a third of the maximum compressive strength.

The early age modulus (3 days) provides key data for massive elements when submitted to THM effects, which are very significant at an early age. However, it is not possible to measure the in-service Young's modulus

1 A duration of 28 days represents the standard and contractual curing phase. For specific concretes, the physical hardening phase can often be extended to 90 days.

2 To simulate the in-service effects on the resistance of the concrete, such as for power-plant water-cooling towers or marine works (e.g. walls), the concrete can also be subject to cyclic wetting and drying.

directly on the test block, so tests are performed on cylindrical concrete specimens (Ø = 16 cm, h = 32 cm) cast at the same time as the mock-up and subject to the same environmental conditions (i.e. kept adjacent to the mock-up). This approach allows verification of any change in Young's modulus, for durations between a few days to several months, depending on the number of sample specimens available.

Additionally, with 3D effects being significant for thick structures, it may be necessary to measure the elasticity modulus values obtained on cylindrical concrete test specimens (Ø = 16 cm, h = 32 cm). These values benefit from the measurements made on thick structures and are used to update the numerical models, adjusting the results obtained by these simulations. Therefore it is necessary to verify all the mechanical characteristics, including the compressive and tensile limit stress, and the Young's modulus of the reinforcement bars used.

#### 11.2.1.4. *Recommendations for the measurement of dilatation and shrinkage*

The sensor measurement is influenced by the concrete thermal behavior "expansion/contraction". Note, the linear concrete thermal expansion coefficient is not constant from early age. The thermal expansion coefficient change depends on the concrete moisture content evolution. Adjustment of the concrete thermal expansion coefficient with the steel thermal expansion is necessary when correcting the sensor measurements.

The free thermal expansion of the hardened concrete depends on: grain size, sand quality, aggregate type, concrete porosity and the quantity of mobile free water present in the concrete pores. Depending on the aggregate characteristics, the concrete linear thermal expansion coefficient can vary from 9 to 14 µm/m $(°C)^{-1}$ during the first months following concrete casting.

For information, the CEOS.fr project experience is that the concrete thermal expansion coefficient obtained using Vibrating Wire Strain Gauges (VWSG) during the 21-day maturing period is in the range of 11 to 12.5 µm/m/°C for the free tests and QAB samples. Hence, the corresponding thermal expansion coefficient is very similar to that used for the steel standard thermal expansion coefficient of 11.5 µm / m / °C. In this scenario, expansion of the steel reinforcement is taken to be the main contributor to the strains measured on the concrete reinforced structures.

NOTE 11.1.– The measured variation is narrower than that given in MC2010 and these guidelines, sections 7–7.6.3, which give a range between $6 \cdot 10^{-6}$ $K^{-1}$ and $15 \cdot 10^{-6}$ $K^{-1}$ .This difference is due to the concrete range considered in MC2010 being wider than the range tested under the CEOS.fr project.

*Concrete specimen global shrinkage estimate (mock-up and structure)*

The use of small size sensors allows a local strain to be obtained and as a result the local shrinkage (in µm/m). As this shrinkage is not homogeneous in all the parts of the test specimen, especially during the maturing phase, then an adequate number of sensors must be located on the test specimen, in order to estimate the local shrinkage at the core and on the surface. The VWSG temperature measurement enables the temperature contribution from shrinkage to be considered separately, hence the thermal expansion can be taken into account, at least qualitatively.

At the end of the maturing phase, the whole test specimen (in the core, along the reinforcement and on the surface) presents an approximately homogeneous shrinkage, although the reinforcement layer and its thickness have an impact on the shrinkage value.

For example, in the case of the CEOS.fr block RL6 after two months of maturing:

– the shrinkage at the end of the maturing phase converges to –80 µm/m adjacent to the rebars;

– in the block core, the measured shrinkage is limited to –55 µm/m, which demonstrates the non-homogeneous behavior obtained for a thick concrete structure of 1 m × 0.8 m cross section.

Given that the shrinkage measured on the concrete specimen commonly called the "free sample test" (Ø 16 cm × 32 cm) described below is –110 µm/m, the adjacent rebar values are consistent.

In the specimen core, it can be assumed that shrinkage is constrained and hence the concrete is subject to local tension.

The complete analysis of all the test specimen measurements enables identification of the separate various phenomena, among which include:

– autogenous shrinkage;

– drying shrinkage;

– basic creep.

The concrete basic creep and autogenous shrinkage should be studied in a laboratory in accordance with well-defined procedures (for example, cylindrical concrete specimens kept at 20°C and 95% relative humidity (RH)).

Interpretation of the micro strains data, obtained from cylindrical concrete specimens (Ø= 16 cm, h = 32 cm) cast on the construction site, requires additional information, measured using instrumentation, which includes vibrating wire extensometers (WVSG, temperature measurement and micro strains, ANDRA's method). The specimens considered are described as follows:

– 16 × 32 cylindrical test specimens stored on the construction site (i.e. subject to the same environmental conditions): a "free specimen" provides free deformation of the concrete mix used, taking account of the effect of the concrete hydration under the same environmental conditions;

– 16 × 32 cylindrical test specimens placed in an insulated box, also stored on the construction site: a "QAB specimen" (semi-adiabatic conditions), which behaves similarly to the concrete in the core of the massive test block.

A "free specimen" gives a representative measurement of the large block specimen concrete surface, without rebar. The QAB specimen measures the endogenous shrinkage, which is increased by the thermal effects linked to the exothermic gradients occurring during the concrete maturing phase. The QAB specimen demonstrates almost the same behavior as the core concrete, which is also free of any rebar effects. The associated data, which is heavily dependent on the casting and environmental conditions (e.g. humidity ratio, ambient temperature, exposure to sunlight for the "free specimen", etc.), supplies relevant and individual limits for every test structures. Hence, analysis of all the results is advisable, especially during application of the mechanical load taking into account these initial conditions.

The values obtained from the QAB specimen tests provide indications of the concrete state at the core of the concrete structure but cannot be interpreted as representative of the autogenous shrinkage, which can only be obtained by appropriate laboratory tests.

Finally, to obtain the complete stress state, the designer also has to take into account the change with time of the concrete Young's modulus in tension and compression. Temperature, hygrometry and associated shrinkage are representative of the concrete state at any given moment.

#### 11.2.1.5. *The cracking pattern: spacing and widths*

The design of concrete massive structures requires both an evaluation of the structural resistance and an evaluation of the cracking pattern under the mechanical loads, including crack widths and spacing between cracks. For many studies, the average crack spacing in stabilized configurations is recorded more often than the crack widths. This is due to the inherent difficulty associated with measuring the crack widths, which vary significantly in width and direction. Assessment of the experimental results raises questions regarding the validity of characteristic width limits defined in the engineering codes. Generally, the maximum crack widths or their average widths are measured.

The crack width measurement is obtained from either:

– local measurement, obtained using optical microscopy or extensometers located on a specific crack point (this location remains subjective);

– or using a standard long base extensometer to measure the extension on a part of the test body which includes the cracks. In this case, width measurement is obtained by dividing the global extension by the number of cracks; this gives an average crack opening value, which also includes the tensile concrete strains.

Comparison of the crack widths between these different methods is particularly difficult given their discrepancies, as discussed in [PER 13].

Use of an automatic measurement (for example using long base tensile wire) at a given point of the test body chosen in advance for each crack, corresponding to the loaded zone, may reduce the uncertainties of the measurement and provide a relationship representative of the crack width change with time and load. A crack, as defined in the engineering codes (for example EC2 & MC2010), corresponds to a visible crack, which is normally perpendicular to the reinforcement layer for the load considered and generally regular until the crack reaches the rebar. In the case of a thick structure, the surface crack width is heavily dependent on the coverage – see Figure 11.2, in accordance with [BOR 10].

Analysis by digital image correlation (DIC) provides a relatively accurate measurement of the visible cracks for every load step, situated on the top of the beam surface. It allows measurement of the initial crack formation and its propagation [RUO 12, ROS 14].

The use of DIC enables any change in the crack to be captured during the application of the mechanical load, providing accurate local information on the crack spacing and crack widths.

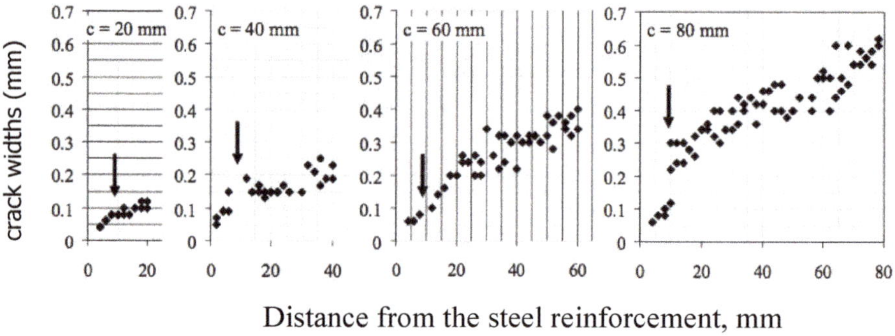


**Figure 11.2.** *Variation of the crack widths obtained for tests with several covers [BOR 10]*

This measurement method uses image correlation, which enables the crack development to be followed throughout the loading process, providing realistic possibilities of data mining, which is much less subjective than the manual recorded measurements generally carried out.

The following are recommendations on the use of DIC methods:

– scatter plots must be produced to create a recognizable and located area during image processing and a means of increasing the contrast with natural concrete, for example without creating too strong a gradient from only the presence of black and white spots;

– scatter plots must be randomly chosen, as a regular pattern makes the recognition of a local specific plot more difficult;

– targets must be placed in a fixed frame around the mock-up and on the mock-up itself. This provides a fixed frame of reference and enables to locate all views of the test specimen relative to this absolute reference;

– for large-sized reduced scale tests the movement of the test specimens can be several centimeters, which makes the accurate repositioning of every image and elimination of rigid body motion essential;

– finally, corrections required for any intensity changes in light are assessed if the test specimen is located externally.

Numerous software packages exist and offer various levels of accuracy in the assessment of the local strain, or alternatively for the measurement of global strain (linked to the overall movement and test specimen displacements). Finally, analysis will have to take into account the effects of the ambient temperature and the hygrometry, which contribute to the concrete strain and hence potentially to crack width development over time.

#### 11.2.1.6. *Shrinkage and creep*

In the selection of the instrumentations, particularly sensors, which estimate the characteristics of concrete shrinkage and creep, it is necessary to distinguish between measurements obtained on concrete:

– cylindrical concrete specimens (Ø = 16 cm, h = 32 cm),

– and structures.

Assessment of the structure's function, and hence the anticipated shrinkage and associated strains generated during the shrinkage, can only be performed on test specimens; see section 11.2.1.4.

### 11.2.2. *Data acquisition and storage*

The automatic data acquisition system allows the management and measurement of various types of sensors, which use various technologies. A time sequenced storage database facilitates the post data acquisition processing. External data processing on computer spreadsheets allows curves to be produced, as function of time, enabling comparison with the change in various parameters, including strain/temperature.

The storage of the data is performed on a database, as defined beforehand, in accordance with its use and expected measurement data-mining.

## 11.3. Measurement of structures

The measurement principles (choice of instrumentation and location) are adapted to either a pre-stressed structure or reinforced concrete structure.

Post selection, the CEOS.fr project used the same monitoring approach, adopting the same technology choice, metrological characteristics and identical suppliers to those typically used for the monitoring of major French civil works.

### *Concrete pre-stressed structures*

A pre-stressed structure, such as a water tank, is normally able to freely experience thermal expansion, avoiding restraint to the pre-stressed loads. The use of implanted short length-based extensometers embedded in the concrete structure is advised. For massive structures, it is recommended that extensometers, applied in three directions, be used. These are located at two or three points in the structure (two close to the surface layer and one at the core of the wall). The three directions chosen consist of two directions parallel to the wall surface, and one perpendicular to it. This approach allows strain measurement in the direction of the principal stress and also a measurement in at least one direction, which is not subject to load. Measurements perpendicular to the outer surface allow measurement of the effects of desiccation and delayed strains between the surface and the core of the works. Prior to pre-stressing it is recommended that the state of the structure and strain level expected due to shrinkage are defined. This approach helps to distinguish the shrinkage phenomena from those applied due to pre-stressing, as described in section 11.2.1.4 "Free specimen and QAB specimen". Each of these tests is instrumented with vibrating wire extensometers which are used to measure:

– the strain of the autogenous shrinkage and drying shrinkage during the first months of the structure's lifetime (QAB specimen);

– the effect of the concrete hydration on the environmental conditions of the structure (i.e. using the free specimen).

It is essential for measurements obtained using extensometers that temperature measurements taken from the same location at the same time are used; this enables the temperature effects on the strains to be discounted. With these measurements being taken in three directions and located within several points of the structure's thickness, it is possible, after thermal correction, to estimate the change in strain with time according to the load conditions and then, in various directions, to estimate the part of strain due to shrinkage and creep.

If the measurements are implemented from the start of concrete casting, it is important to avoid weak points in the structure which could be sensitive to early age cracking before pre-stressing settlement (i.e. locations near pre-stress tendons, possible grooves or bending zones). For such a pre-stressed structure, displacement measurement or long base extensometers provide local measurements, but these can be difficult to interpret and require a large number of temperature measurements to reliably eliminate thermal strains.

In the case of structures working mainly in bending (i.e. mainly subjected to bending forces), such as bridges, global displacement measurements may be more accurate than local measurements, which are too sensitive to the load and moment distribution. Use of an appropriate number of temperature measurements allows the thermal strain contribution to be discounted. Trend curves must be used where reliable thermal corrections are lacking.

### *Reinforced concrete structures*

Reinforced concrete structures are more sensitive to cracking than pre-stressed structures and are generally more hyperstatic. Therefore, interpretation of local strains is more difficult because only restrained strains can be measured.

In this case, it is necessary to use (see section 11.2.1.4 "Free specimens and QAB specimens") each specimen instrumented by a VWSG to measure:

– the strain due to autogenous and drying shrinkage and during the structural lifecycle conditions (QAB);

– the effect of concrete hydration on the environmental conditions of the structure.

Similar to pre-stressed structures, it is necessary to take extensometer and temperature measurements from the same location, at the same time, enabling thermal strains to be discounted.

The differences between measurements from VWSG embedded in the concrete of the structure and those included in test specimens (free or QAB) allow the strains to be estimated. As for pre-stressed works, the interpretation of local measurements for bending deformation can be a more difficult process to apply.

For highly hyperstatic reinforced concrete structures (for example concrete dams), global displacement measurements can be more accurate than local measurements, which are too sensitive to stress and moment distribution. A large number of temperature measurements are necessary to enable the separation of thermal strains. Trend curves must be used where reliable thermal corrections are lacking.

### 11.3.1. *Preliminary measurements*

Firstly, the specific data, which characterizes the structure materials, is confirmed. Use of an automatic meteorological facility (station) is recommended to identify the parameters which have an influence on the behavior of massive structures, such as external temperatures measured in various locations near the structure, wind speed and orientation, sun effects, and the associated barometric pressure, etc. Synchronization of the data acquisition provides a basis for common measurement. It is strongly recommended that all the collected data is stored in a time-sequenced manner.

### 11.3.2. *Parameters to be measured*

The project manager and/or his technical advisers, for example designers, shall determine and select the essential parameters that have to be monitored to ensure both the structure's short-term durability and serviceability, taking into account its environmental design conditions and in-service requirements. Hence, the chosen parameters should provide suitably accurate measurements to be used in the numerical calculations "modeling and predictive standards of structure behavior". These parameters constitute the basis of the technical specification for the measurement system.

### 11.3.3. *Equipment of the measurements*

#### 11.3.3.1. *Implementation of the temperature probes and strain sensors*

Following selection of the most suitable temperature sensors for the reinforced concrete works under consideration, the requirements of Part 65 section 8.3.2.5.5.1.1 [SET 14] are applied, and the measured results compared with the theoretical expectations of the temperature distribution in the analysis.

It should be noted that the test of suitability requires the preparation of a concrete specimen and concerns the study of at least a "representative section of elements to be thermally handled".

The same approach is adopted for sensor selection and their positioning, comparing the extensometer micro-strain measured result predictions.

For internal strains, VWSGs are embedded in various locations of the structure, including the core of the largest central part of the structure and distributed on the surface.

#### 11.3.3.2. *Free test specimen and Quasi-Adiabatic Box (QAB) specimen*

The data analysis obtained from the VWSG and the structure's temperature sensors is compared with the behavior of the free test specimen and QAB cylindrical concrete specimen quasi-adiabatic condition; see section 11.2.1.4.

Two test specimens, one free strain exposed to environmental conditions and one QAB specimen, must be cast at the same time as the structure, including its associated VWSG:

– the "free test specimen": (cylindrical Ø 16 cm, h = 32 cm) is instrumented with a VWSG and located adjacent to the structure, and as a result it is then subject to the same structural environmental conditions albeit protected from the rain;

– the "QAB specimen": (cylindrical Ø = 16 cm, h = 32 cm) is placed in a QAB, located near the structure. The specimen is used to measure the autogenous shrinkage, which increases the thermal effect as a result of the link to the exothermic gradients experienced during the maturing of the concrete in the QAB.

The values acquired by the “free test specimen” VWSG are representative of the concrete behavior at the concrete surface of the works. This test measurement gives the temperature change at an early age and a shrinkage strain, which is different from that provided by the QAB specimen VWSG. The values given by the QAB specimen VWSG sensor are more representative of the concrete core of a massive structure.

From data supplied by both test specimens, interpretation of the measurements made on the structure can be performed and a numerical simulation made to determine the mechanical concrete strains in the reinforced concrete of the massive structures.

### 11.3.4. *Formwork*

The formwork is the component that participates in the thermal exchange with the external environment. Hence, the peak concrete temperature experienced in contact and in the works’ mass during the maturing phase is influenced by this thermal exchange. Formwork should be insulated to avoid thermal shocks.

## 11.4. Example of measurement instrumentation on massive structures

Numerous examples of the instrumentation used on civil engineering massive structures are available following agreement with the owner of these works.

The measurement systems normally used in the monitoring of reinforced concrete (RC) structures almost always uses the same technology as that used in the CEOS.fr project for reduced scale tests. Operators and/or data acquisition systems capture the measurements acquired.

As a result of the large scale of the massive structures, it is necessary to comply with security rules and relevant operational legislation in force. However, expert assessment is still required, incorporating information from in-service monitoring measurements (such as sensor based instrumentation).

"Expert assessment is still considered to be the most reliable approach for evaluating and validating that measurements conform to specified requirements in accordance with the owner's quality programme".

Operators for all type of works use the following methods:

– visual inspections by a qualified operator, using automatic methods such as "pictures, photogrammetry, scanner, etc.";

– "topographic" campaigns by a surveyor, using automatic methods such as "motorised theodolite, scanner, drone, etc.";

– the formal reports produced by these experts are integrated into the database of the work and for each assessment include: the activity date, all associated relevant environmental conditions and information necessary for a good understanding of any locally acquired differences (for example: local temperature, water level, etc.).

## 11.5. Example of mock-up test instrumentation

The CEOS.fr project was divided into three parts, based on the load considered, namely: monotonous static loads, thermo-hydro-mechanical loads and cyclic loads, leading to the use of three specific test bodies.

For each type of test some common principles were identified:

– cast concrete test bodies are to be as large as permitted by the test machine or by physical constraints;

– design test bodies are to be as geometrically simple as possible in order to allow a representative modeling;

– starting with one reference test body, then making just one "second order" parameter change, allows checking of the calculation model, confirming it is able to reliably represent even slight changes in detail which are likely to influence crack generation;

– given it is not possible to effectively predict where and how cracks will appear, it has been decided to use numerous sensors which are based on different physical principles. Hence, it is anticipated that relevant information can be captured from at least one set of available data;

– relevant material characteristics are to be determined thoroughly (based on no less than 15 parameters for concrete) and from an early age for concrete strength and modulus.

Given the difficulties associated with ensuring stable ambient conditions, a complete set of additional sensors (wind and sun effects, temperature and hygrometry) are to be placed close to the test body structure in the test area. These sensors will provide a full set of boundary conditions, which can then be used in the numerical modeling.

The detailed plan and description of the experimental test procedures are archived.

A series of tests were carried out on instrumented prismatic test blocks (RL and RG), third-scale walls and third-scale prismatic blocks (RL 1/3). Each test provides data to support the standard calculations made using finite element and predictive crack modeling. The most instrumented test bodies are the prismatic blocks (RL), which are maturated freely. These blocks are fully instrumented with all the available instrumentation and a complete assessment made, with cross comparisons performed.

Figure 11.2 below presents the zone of interest (constant moment zone = center part of the beam) giving the location of the main sensors and measurement techniques:

– upper surface displacement measurements are given by linear variable differential transformer (LVDT), which allows verification of the global deformation based on the beam displacement;

– long length base optical extensometers are used, located both on an external face and embedded inside the core of the specimen;

– VWSGs are embedded in the core of the specimen (identified with CV in Figure 11.2);

– optical and temperature sensors are incorporated using Fiber Bragg Grating (FBG);

– strain gauges are located on the reinforcement layers;

– roughcast done for the DIC used to define the crack location reconstructed in the last step of load.

The DIC used for all large RL beams and shear-walls, provides a cracking pattern resolution of 0.05 mm for blocks and test walls.

In addition to the mock-up instrumentation described above, the following are also used in the tests:

– temperature measurement of distributed Pt100 probes;

– local strain measurement (VWSG) – microdefs – obtained on cylindrical test specimens; cylindrical Ø 16 cm, h = 32 cm free specimen and QAB specimens.

The interpretation of early age concrete was mainly made from local temperature measurements and comparison with two specimens (free and QAB) each in two different thermal and moisture situations (free – local environment and QAB – almost adiabatic) and beam measurements. Hence, it is possible to identify the test block temperature gradients and their change during the exothermic phase, and subsequently during the maturing phase. The temperature gradients can be used to evaluate whether any early damage took place; this is generally characterized by loss of the VWSG signals, or an abnormal increase in ambient thermal oscillations (day–night). As a result of surface cracking, it is possible to confirm the stress redistribution between concrete and rebar.

The changing crack pattern is checked from early age and the occurrence of crack closure during the maturing phase, under the influence of shrinkage, is observed. As a result, the global shrinkage and also all the beam history during application of the bending load moment are known.

During the mechanical load, instrumentation, especially long length base optical fiber, provides global data about the displacement. Note, that the occurrence of damage in zones near the VWSG renders the corresponding measurements unfit for use.

NOTE 11.2.– Special care should be taken when considering early age concrete that contributes to the behavior of the structure during its operational lifetime.

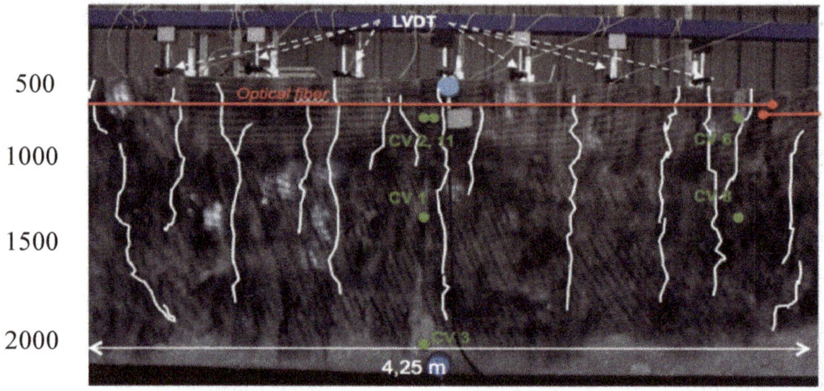


**Figure 11.3.** *Measurement on a massive beam, in the central zone of the constant moment*

## 11.6. Conclusion

The structural survey and instrumentation of special massive structures, and the possible use of mock-up scale test facilities, are essential operations that extend the knowledge and enhance the understanding of the short-term and long-term behavior of such special structures.

Once integrated into a numerical process comparing experience with feedback and real time mechanical behavior simulation, the acquired data are a key tool in the decision-making process for design and operational management. For the owner, it is also a source of data that can be used to assess the work's serviceability and safety, where accurate indicators based on measurements have been previously defined. Finally, this approach supports the organization, which is responsible for confirming that the structure performance is in accordance with theoretical expectations.

# Bibliography

[ACI 07] ACI, Effect of restraint, volume change and reinforcement on cracking of mass concrete, report, American Concrete Institute, 2007.

[ACK 88] ACKER P., Comportement mécanique des bétons: approche physico-chimique – Rapport de recherche LCP no. 152, LCPC/IFSTTAR. LCPC, 1988.

[ARI 06] ARIOGLU N., GIRGIN C., ARIOGLU E., "Evaluation of ratio between splitting tensile strength and compressive strength for concretes up to 120 MPa and its application in strength criterion", *ACI Materials Journal*, vol. 103, no. M03, 2006.

[BAE 13] BAE S., "Alternative design approach for thermal effects", *22nd International Congerence on Structural Mechanics in Reactor Technology,* San Francisco, 2013.

[BAL 93] BALAZS G., "Cracking analysis based on slips and bonds stresses", *ACI Materials Journal*, vol. 90, no. 4, pp. 340–348, 1993.

[BAR 07] BARRÉ F., CHATAIGNER J., CHAUVEL D., "Crack width prediction in thick concrete elements using EC2 code", *FIB symposium*, Dubrovnik, 2007.

[BAR 14] BARRÉ F., Crack control at an early age and long term operation, CEOS Concrack4, March 2014.

[BAR 99] BAROGHEL-BOUNY V., MAINGUY M., LASSABATERE T. *et al.*, "Characterization and identification of equilibrium and transfer moisture properties for ordinary and high-performance cementious materials", *Cement and Concrete Research*, vol. 29, no. 8, pp. 1225–1238, 1999.

[BAZ 01] BAZANT Z., "Prediction of concrete creep and shrinkage: past, present and future", *Nuclear Engineering and Design*, vol. 203, no. 1, pp. 27–38, 2001.

[BAZ 04] BAZANT Z., "Probability distribution of energetic-statistical size effect in quasibrittle fracture", *Probabilistic Engineering Mechanics*, 2004.

[BAZ 05] BAZANT Z., YAVARI A., "Is the cause of size effect on structural strength fractal or energetic statistical?", *Engineering Fracture Mechanics*, vol. 72, pp. 1–31, 2005.

[BEN 05] BENBOUDJEMA F., MEFTAH F., TORRENTI J., "Interaction between drying shrinkage creep and cracking phenomena in concrete", *Engineering Structures*, vol. 27, no. 2, pp. 259–250, 2005.

[BOR 10] BOROSNYOI A., SNOBLI I., "Crack width variation within the concrete cover of reinforced concrete members", *Journal of Hungarian Scientific Society of the Silicate Industry*, January 2010.

[BUF 11] BUFFO-LACARRIÈRE L., SELLIER A., TURATSINZE A. *et al.*, "Finite element modelling of hardening concrete: application to the prediction of early age cracking for massive reinforced structures", *Materials and Structures*, vol. 44, no. 10, pp. 1821–1835, 2011.

[BUF 14] BUFFO-LACARRIÈRE L., SELLIER A., KOLANI B., "Application of thermo-hydro-chemico-mechanical model for early age behaviour of concrete to experimental massive reinforced structures with strain-restraining system", *European Journal of Environmental and Civil Engineering*, vol. 18, no. 7, pp. 814–827, 2014.

[BUF 16] BUFFO-LACARRIÈRE L., BARON S., BARRÉ F. *et al.*, "Restrained shrinkage of massive reinforced concrete structures: results of the project CEOS.fr", *European Journal of Environmental and Civil Engineering*, vol. 20, pp. 785–808, 2016.

[CAR 95] CARPINTERI A., CHIAIA B., FERRO G., "Size effect on nominal tensile strength of concrete structures: multifractability of material ligament and dimensional transition from order to disorder", *Mater Structure*, vol. 28, pp. 311–317, 1995.

[CEB 12] CEB-FIP, "Model Code 2010 final draft", Bulletin fib, pp. 65–66, March 2012.

[CEB 93] CEB-FIP, "Model Code 90", CEB Bulletin, no. 213/214, 213/214, 1993.

[CHA 06] CHAUVEL D., TOURET J.-P., BARRÉ F., "Assessment of long term concrete deformations of nuclear structures based on in situ measurements", *First Euro Mediterranean Symposuim in Advances on Geomaterials and Structures*, Hammamet, Tunisia, 2006.

[CHA 07] CHAUVEL D., BARRÉ F., "Thermal and mechanical behaviour of an experimental mock-up of a nuclear containement", *CONSEC 07*, Tours, 2007.

[CIR 07] CIRIA-UK, Early-age thermal crack control in concrete, UK Guide C660, CIRIA, London, 2007.

[DEB 10] DEBERNARDI P., TALIANO M., "Cracking of pre-tensioned structures", *3rd Fib International Congress*, June 2010.

[DEB 13] DEBERNARDI P., GUIGLIA M., TALIANO M., "Effect of secondary cracks for cracking analysis of reinforced concrete tie", *ACI Materials Journal*, vol. 110, no. M20, 2013.

[ELI 82] ELIGEHAUSEN R., POPOV E., BERTERO V., "Local bond stress-slip relationships of deformed bars under generalize excitations", *Proceedings of the 7th European Conference on Earthquake Engineering*, Athens, vol. 4, pp. 69–80, 1982.

[FAR 95] FARRA B., "Influence de la résistance du béton et de son adhérence avec l'armature sur la fissuration", PhD thesis, 1359, EPFL, 1995.

[GRA 96] GRANGER L., Comportement mécanique des bétons dans les enceintes nucléaires, Analyse et modélisation, LCPC & EDF, PhD thesis, ENPC, 1996.

[ISO 04] ISO 1920-3, Testing of concrete Part 3: Making and curing test ISO, October 2004.

[JCI 08] JCI, "Guidelines for control of cracking mass concrete", Japan Concrete Institute, 2008.

[KAU 98] KAUFMANN W., "Strength and deformations of structural concrete subjected to in-plan shear and normal forces", PhD thesis, Swiss Federal Institute of Technology, Zurich, 1998.

[KOL 12] KOLANI B., BUFFO-LACARRIÈRE L., SELLIER A. *et al.* "Hydration of slag-blended cements", *Cements and Concrete Composites*, vol. 34, no. 9, pp. 1009–1018, 2012.

[KRE 95] KRETZ T., Ponts mixtes recommandations pour maitriser la fissuration des dalles, SETRA, Bagneux, France, 1995.

[LAD 11] LADAOUI W., VIDAL T., SELLIER A. *et al.*, "Effect of a temperature change from 20 to 50°C on the basic creep of HPC and HPFRC", *Mater. Struct.*, vol. 44, no. 9, pp. 1629–1639, 2011.

[LCP 07] LCPC -IFSTTAR., "Recommandations pour la prévention des désordres dus à la réaction sulfatique interne", Guide Technique, 2007.

[LHE 73] L'HERMITE R., "Influence de la dimension absolue sur la résistance à la flexion", *Annales de l'ITBTP*, vol. 30, no. 310, pp. 34–41, 1973.

[MAT 10] MATALLAH M., LA BORDERIE C., MAUREL O., "A practical method to estimate crack openings in concrete structures", *International Journal of Numerical and Analytical Methods in Geomechanics*, vol. 34, no. 15, pp. 1615–1633, 2010.

[MAZ 86] MAZARS J., "A description of micro and macroscale damage of concrete structures", *Engineering Fracture Mechanics*, vol. 25, nos. 5–6, pp. 729–737, 1986.

[MEN 88] MENSI R., ACKER P., ATTOLOU A., "Séchage du béton: analyse et modélisation", *Materials and Structures*, vol. 21, no. 1, pp. 3–12, 1988.

[MER 14] MERLE P., Essais de caractérisation de la fissuration des tirants comportant plusieurs armatures longitudinales et soumis à traction simple, EDF SEPTEN. IREX, 2014.

[MET 15] METEO FRANCE, Annual Climate Report, Meteo France, 2015.

[NF 04] NF EN 1992-1-1, Eurocode 2-1 – design of concrete structures, Part 1-1 – general rules for buildings, AFNOR, 2004.

[NF 05a] NF P 18-711-1/NA, Eurocode 2-1 annexe nationale, AFNOR, 2005.

[NF 05b] NF P 18-721-1/NA, Eurocode 2-2 annexe nationale, AFNOR, 2005.

[NF 06a] NF EN 1992-2, Eurocode 2-2 – design of concrete structures Part 2 – concrete bridges design and detailing rules, AFNOR, 2006.

[NF 06b] NF EN 1992-3, Eurocode 2-3 – design of concrete structures, Part 3: liquid retaining and containment structures, AFNOR, 2006.

[NF 13] NF EN 13670, Execution of concrete structures, AFNOR, 2013.

[OLI 15] OLIVIER-LEBLOND C., DELAPLACE A., RAGUENAU F., "Modeling of three-dimensional crack patterns in deep reinforced concrete structures", *Engineering Structures*, vol. 83, pp. 176–9186, 2015.

[PER 13] PEREZ-CALDENTEY A., CORRES-PEIRETTI H., PESET-IRIBARREN J. *et al.*, "Cracking of RC members revisited: influence of cover, φ/ρs, *ef* and stirrup spacing – an experimental and theoretical study", *Structural Concrete*, vol. 14, no. 1, pp. 69–78, 2013.

[PIJ 87] PIJAUDIER-CABOT G., BAZANT Z., "Nonlocal damage theory", *Journal of Engineering Mechanics*, vol. 113, pp. 1512–1533, 1987.

[PN 13a] PN CEOS.FR, LP1 – Synthèse des résultats expérimentaux, 2013.

[PN 13b] PN CEOS.FR, LP3 – Résultats obtenus dans la connaissance des phénomènes de fissuration, Livrable 3, IREX, Paris, 2013.

[PN 14] PN CEOS.FR, LP2 – Récapitulation des résultats, Livrable principal 2, IREX, 2014.

[RIL 12] RILEM – CEOS.FR – EFB, Concrack3, Concrack3 workshop, IFSTTAR, 2012.

[ROM 11] ROMANIAN STANDARD INSTITUTE (RSI), Guide de conception pour le contrôle de la fissuration des éléments lourds et des murs porteurs en béton armé due au retrait empêché, Norme GP-115, Romania, 2011.

[ROS 12] ROSSI P., TAILHAN J., "Cracking of concrete structures: interest and advantage of the probabilistic discrete approaches", *SSCS Conference*, Aix en Provence, 2012.

[ROS 13] ROSPARS C., Courbes expérimentales relatives aux poutres RL, CEOS.fr IREX, 2013.

[ROS 14] ROSPARS C., CHAUVEL D., "CEOS.fr experimental programme and reference specimen tests results", *European Journal of Environemental and Civil Engineering*, vol. 18, no. 7 Special Issue, pp. 738–753, 2014.

[ROS 94] ROSSI P., WU X., LE MAOU F. *et al.*, "Scale effect on concrete in tension", *Materials and Structures*, vol. 27, no. 8, pp. 437–444, 1994.

[RUO 12] RUOCCI G., "Cracks distance and opening in reinforced concrete membranes: experimental results from monotonic and cyclic loading histories", *15th World Conference on Earthquake Engineering*, 2012.

[SEL 09] SELLIER A., BUFFO-LACARRIÈRE, L., "Towards a simple and unified modelling of basic creep, shrinkage and drying creep of concrete", *European Journal of Environmental and Civil Engineering*, vol. 13, no. 10, pp. 1161–1182, 2009.

[SEL 12a] SELLIER A., Modélisation probabiliste pour un chargement THM – Prise en compte des effets d'échelle aléatoires induits par le retrait différentiel pâte-granulat, rapport de recherche, ANRT, Groupe de travail 4 du projet de recherche ANRT Mefisto, Toulouse, 2012.

[SEL 12b] SELLIER A., BUFFO-LACARRIÈRE L., MULTON S. *et al.*, "Nonlinear basic creep and drying creep modeling", in ROSSI T. (ed.), *SSCS conference*, Aix en Provence, 2012.

[SEL 12c] SELLIER A., BUFFO-LACARRIERE L., SOUYRIS P. *et al.*, Rapport final de développement des modèles THM – Contribution du LMDC – Groupe de travail 2 du projet de recherche ANRT Mefisto, Recherche Mefisto, ANRT, Toulouse, 2012.

[SEL 13a] SELLIER A., MILLARD A., "WL2, une méthode pour considérer l'effet d'échelle probabiliste dans les modèles de fissuration", Rencontres AUGC, ENS de Cachan, 2013.

[SEL 13b] SELLIER A., CASAUX-GINESTET G., BUFFO-LACARRIÈRE L. *et al.*, "Orthotropic damage coupled with localised crack reclosure processing, Part I: Constitutive Laws", *Engineering Fracture Mechanics*, vol. 971, pp. 148–167, 2013.

[SEL 14a] SELLIER A., MILLARD A., "Dépendances spatio-temporelles de la résistance des matériaux. Modélisation pour une prise en compte en contexte non linéaire", *8ème journée de fiabilité des matériaux et des structures*, Aix en Provence, 2014.

[SEL 14b] SELLIER A., MILLARD A., "Weakest link and localisation WL 2: a method to conciliate probabilistic and energetic scale effects in numerical models", *European Journal of Environmental and Civil Engineering*, vol. 18, no. 10, pp. 1177–1191, 2014.

[SEL 16] SELLIER A., MULTON S., BUFFO-LACARRIÈRE L. *et al.* "Concrete creep modelling for structural applications: non-linearity, multi-axiality, hydration, temperature and drying effects", *Cement and Concrete Research*, vol. 79, pp. 301–315, 2016.

[SET 14] SETRA – MINISTÈRE DE l'ÉCOLOGIE DU DÉVELOPPMENT DURABLE ET DE L'ÉNERGIE, Cahier des Clauses Techniques générales. Fascicule 65 – Exécution des ouvrages de génie civil en béton, Paris, 2014.

[SPO 10] SPOIDEN A., Rédaction des références bibliographiques selon les normes de l'American Psychological Association (APA), Université Catholique de Louvain, Bibliothèque de Psychologie et des Sciences de l'éducation, 2010.

[VAN 00] VAN VLIET M., VAN MIER J., "Experimental investigation of size effect in concrete and sandstone under uniaxial tension", *Engineering Fracture Mechanics*, vol. 65, pp. 165–188, 2000.

[VAN 85] VAN GENUCHTEN T.M., NIELSEN D., "On describing and predicting the hydraulic properties of unsaturated soils", *Annales Physicae*, vol. 3, no. 5, pp. 615–628, 1985.

[VEC 86] VECCHIO F.J., COLLINS M.P., "The modified compression-field theory for reinforced concrete elements subjected to shear", *ACI Journal*, vol. 83, no. 2, pp. 219–231, 1986.

[YAM 14] YAMAMOTO Y., NAKAMURA H., KURODA I. *et al.*, "Crack propagation analysis of reinforced concrete wall under cyclic loading using RBSM", *European Journal of Environmental and Civil Engineering*, vol. 18, no. 7, pp. 780–792, 2014.

# Index

## A, B, C


autogeneous shrinkage, 124, 125, 139, 141, 144, 168, 173, 204, 205, 208, 211
basic creep, 129, 134, 139–141, 178–181, 204
beams, 59
combining effects, 164–169
concrete
  structures, 1, 116, 152, 189, 190, 193, 209, 210
  tensile strength, 33–35, 37, 49, 106, 138
  volume, 16, 32, 33, 35, 36, 163
crack
  angle, 77–79, 81, 83, 104
  initiation, 196
  spacing, 59
  width, 19, 60–68
cracking
  phenomena, 1, 2, 6, 24, 92
  stage, 8, 11, 33, 53, 60, 61, 63, 113–114, 147, 152
creep, 129–133


## D, E, F


data–mining, 194–196
design value, 47, 48
drying
  creep, 131–134, 182, 183
  profile, 171
  shrinkage, 125–129, 141, 142
early age, 27
engineering MC2010, 206
Eurocode 2, 116
Flexural loading, 40, 59


## H, I, M, N


Hydration, 27
  effect, 27
imposed deformation, 32, 115, 116, 124–126, 129, 130, 132, 164–169
instrumentation, 21, 189, 191–193, 196, 204, 207, 212–216
massive
  concrete elements, 119, 120

elements, 8, 9, 16, 21, 32, 36, 51, 59, 115, 120, 143, 146, 164, 189, 201
reinforced concrete, 189, 197
structures, 1, 2, 6, 9, 21, 189, 190, 205, 208, 210, 212, 213, 216
mean differential strain, 50, 51, 82–87
measuring device, 191, 193, 194, 196, 198
mock–up test, 189
modelling, 155, 171
monotonic loading, 10, 21
numerical modelling, 171


## P, R, S


prestressed concrete ties, 111–115
reinforcement ratio, 3, 16, 24, 25, 48, 50, 87, 153, 154
restrained deformations, 143
scale effect, 27
seismic behaviour, 2
shear
force, 8, 13, 75, 77–79, 103, 104
stress, 77, 78, 92, 93, 150, 151, 167
shrinkage, 119
slippage length, 76, 79–82
SLS, 5
stiffness, 61, 92, 114, 147, 148, 150, 152–162, 172, 178, 181
structural monitoring, 191


## T, W


temperature
differential, 27, 198, 199
effect, 135–142, 209
profile, 9
tensile stress–strain curve, 60
testing, 2–20, 26, 138
thermal
effect, 146,152, 153, 204, 211
stresses, 152
thermal–hydro–mechanical effects, 166
thick concrete elements, 105
ties, 47
working loading, 69, 72

Other titles from

ISTE

in

Civil Engineering and Environmental Geomechanics

## 2016

PIJAUDIER-CABOT Gilles *et al.*
*Electrohydraulic Fracturing of Rocks*

TORRENTI Jean-Michel, LA TORRE Francesca
*Materials and Infrastructures 1*
*Materials and Infrastructures 2*

## 2015

AÏT-MOKHTAR Abdelkarim, MILLET Olivier
*Structure Design and Degradation Mechanisms in Coastal Environments*

MONNET Jacques
*In Situ Tests in Geotechnical Engineering*

## 2014

DAÏAN Jean-François
*Equilibrium and Transfer in Porous Media – 3-volume series*
*Equilibrium States – Volume 1*
*Transfer Laws – Volume 2*
*Applications, Isothermal Transport, Coupled Transfers – Volume 3*

## 2013

AMZIANE Sofiane, ARNAUD Laurent
*Bio-aggregate-based Building Materials: Applications to Hemp Concretes*

BONELLI Stéphane
*Erosion in Geomechanics Applied to Dams and Levees*

CASANDJIAN Charles, CHALLAMEL Noël, LANOS Christophe, HELLESLAND Jostein
*Reinforced Concrete Beams, Columns and Frames: Mechanics and Design*

GUÉGUEN Philippe
*Seismic Vulnerability of Structures*

HELLESLAND Jostein, CHALLAMEL Noël, CASANDJIAN Charles, LANOS Christophe
*Reinforced Concrete Beams, Columns and Frames: Section and Slender Member Analysis*

LALOUI Lyesse, DI DONNA Alice
*Energy Geostructures: Innovation in Underground Engineering*

LEGCHENKO Anatoly
*Magnetic Resonance Imaging for Groundwater*

## 2012

BONELLI Stéphane
*Erosion of Geomaterials*

JACOB Bernard *et al.*
*ICWIM6 – Proceedings of the International Conference on Weigh-In-Motion*

OLLIVIER Jean-Pierre, TORRENTI Jean-Marc, CARCASSES Myriam
*Physical Properties of Concrete and Concrete Constituents*

PIJAUDIER-CABOT Gilles, PEREIRA Jean-Michel
*Geomechanics in $CO_2$ Storage Facilities*

## 2011

BAROTH Julien, BREYSSE Denys, SCHOEFS Franck

*Construction Reliability: Safety, Variability and Sustainability*

CREMONA Christian
*Structural Performance: Probability-based Assessment*

HICHER Pierre-Yves
*Multiscales Geomechanics: From Soil to Engineering Projects*

IONESCU Ioan R. *et al.*
*Plasticity of Crystalline Materials: from Dislocations to Continuum*

LOUKILI Ahmed
*Self Compacting Concrete*

MOUTON Yves
*Organic Materials for Sustainable Construction*

NICOT François, LAMBERT Stéphane
*Rockfall Engineering*

PENSE-LHERITIER Anne-Marie
*Formulation*

PIJAUDIER-CABOT Gilles, DUFOUR Frédéric
*Damage Mechanics of Cementitious Materials and Structures*

RADJAI Farhang, DUBOIS Frédéric
*Discrete-element Modeling of Granular Materials*

RESPLENDINO Jacques, TOUTLEMONDE François
*Designing and Building with UHPFRC*

**2010**

ALSHIBLI A. Khalid
*Advances in Computed Tomography for Geomechanics*

BUZAUD Eric, IONESCU Ioan R., VOYIADJIS Georges
*Materials under Extreme Loadings / Application to Penetration and Impact*

LALOUI Lyesse
*Mechanics of Unsaturated Geomechanics*

NOVA Roberto

*Soil Mechanics*

SCHREFLER Bernard, DELAGE Pierre
*Environmental Geomechanics*

TORRENTI Jean-Michel, REYNOUARD Jean-Marie, PIJAUDIER-CABOT Gilles
*Mechanical Behavior of Concrete*

## 2009

AURIAULT Jean-Louis, BOUTIN Claude, GEINDREAU Christian
*Homogenization of Coupled Phenomena in Heterogenous Media*

CAMBOU Bernard, JEAN Michel, RADJAI Fahrang
*Micromechanics of Granular Materials*

MAZARS Jacky, MILLARD Alain
*Dynamic Behavior of Concrete and Seismic Engineering*

NICOT François, WAN Richard
*Micromechanics of Failure in Granular Geomechanics*

## 2008

BETBEDER-MATIBET Jacques
*Seismic Engineering*

CAZACU Oana
*Multiscale Modeling of Heterogenous Materials*

HICHER Pierre-Yves, SHAO Jian-Fu
*Soil and Rock Elastoplasticity*

JACOB Bernard *et al.*
*HVTT 10*

JACOB Bernard *et al.*
*ICWIM 5*

SHAO Jian-Fu, BURLION Nicolas
*GeoProc2008*

## 2006

BALAGEAS Daniel *et al.*
*Structural Health Monitoring*

DESRUES Jacques *et al.*
*Advances in X-ray Tomography for Geomaterials*

FSTT
*Microtunneling and Horizontal Drilling*

MOUTON Yves
*Organic Materials in Civil Engineering*

## 2005

PIJAUDIER-CABOT Gilles *et al.*
*Creep Shrinkage and Durability of Concrete and Concrete Structures CONCREEP – 7*

www.ingramcontent.com/pod-product-compliance
Lightning Source LLC
Chambersburg PA
CBHW071253180926
51613CB00001B/34
*9781786300522*